Nijhoff International Philosophy Series

VOLUME 42

The titles published in this series are listed at the end of this volume.

THEORY OF LANGUAGE SYNTAX

Urszula Wybraniec-Skardowska

Theory of Language Syntax

Categorial Approach

KLUWER ACADEMIC PUBLISHERS
DORDRECHT / BOSTON / LONDON

Library of Congress Cataloging-in-Publication Data

Wybraniec-Skardowska, Urszula.
[Teorie języków syntaktycznie kategorialnych. English]
Theory of language syntax : categorial approach / by Urszula Wybraniec-Skardowska.
p. cm. -- (Nijhoff international philosophy series ; v. 42)
Translation of: Teorie języków syntaktycznie kategorialnych.
Includes bibliographical references and index.
ISBN 0-7923-1142-6
1. Categorial grammar. 2. Formal languages. 3. Logic, Symbolic and mathematical. I. Title. II. Series.
P161.W913 1991
415--dc20 91-64

ISBN 0-7923-1142-6

Published by Kluwer Academic Publishers,
P.O. Box 17, 3300 AA Dordrecht, The Netherlands.

Kluwer Academic Publishers incorporates
the publishing programmes of
D. Reidel, Martinus Nijhoff, Dr W. Junk and MTP Press.

Sold and distributed in the U.S.A. and Canada
by Kluwer Academic Publishers,
101 Philip Drive, Norwell, MA 02061, U.S.A.

In all other countries, sold and distributed
by Kluwer Academic Publishers Group,
P.O. Box 322, 3300 AH Dordrecht, The Netherlands.

Printed on acid-free paper

This book has originally been published in Polish, entitled
Theorie Języków Syntaktycznie Kategorialnyck
with PWN, Warsaw, 1985.

Translated from the Polish by Olgierd Adrian Wojtasiewicz

Printed in the Netherlands

To the memory of Jerzy Słupecki

CONTENTS

Introduction ix

1. Main assumptions, objectives, conditionings ix

2. Intuitive foundations xv

Chapter I

The Axiomatic Theory TLTk of Label Tokens 1

Sec.I.1. Primitive concepts 1

Sec.I.2. Label tokens and their equiformity 3

Sec.I.3. Concatenation 5

Sec.I.4. Vocabulary 10

Sec.I.5. Word tokens 12

Sec.I.6. *n*-componential words; length of word 17

Sec.I.7. Generalized concatenation 23

Sec.I.8. Methodological remarks 29

Chapter II

The Axiomatic System TSCL of Simple Categorial Languages . 32

Sec.II.1. The principal objectives of construction the theory of syntactically categorial languages. 32

Sec.II.2. Connections between TLTk and TSCL 37

Sec.II.3. Categorial indices and their indication 42

Sec.II.4. Expressions 46

Sec.II.5. Well-formed expressions 51

Sec.II.6. Syntactic categories 61

Sec.II.7. The fundamental theorems of the theory of syntactic categories 66

Sec.II.8. The algorithm of checking the syntactic correctness of expressions 78

Chapter III

The Theory **TSCω-L** of Categorial ω-languages 90

Sec.III.1. Introductory remarks 90

Sec.III.2. The foundations of **TSCω-L** 95

Sec.III.3. Operator expressions 97

Sec.III.4. Well-formed expressions 103

Sec.III.5. Fundamental theorems 109

Sec.III.6. The algorithm of checking the syntactic correctness of expressions 120

Chapter IV

Dual Theories 132

Sec.IV.1. The double ontological character of linguistic objects and the biaspectual approach to language 132

Sec.IV.2. The theory **TLTp** of label types 139

Sec.IV.3. Interpretation in **TLTp** of Tarski's axioms of metascience 148

Sec.IV.4. The theory **TETp** of expression types 151

Sec.IV.5. The dual theory **DTSCL** 164

Sec.IV.6. The dual theory **DTSCω-L** 171

Final Remarks 181

Annex 186

Notes 233

References 237

INTRODUCTION

1. Main assumptions, objectives and conditionings

1.1. The present book is concerned with certain problems in the logical philosophy of language. It is written in the spirit of the Polish logical, philosophical, and semiotic tradition, and shows two conceptions of the syntax of *categorial* languages: the theory of ***simple languages***, i.e., languages which do not include variables nor the operators that bind them (for instance, large fragments of natural languages, languages of well-known sentential calculi, the language of Aristotle's traditional syllogistic, languages of equationally definable algebras), and the theory of ***ω-languages***, i.e., languages which include operators and variables bound by the latter (for instance, the language of the functional calculus and languages of the theories based on that calculus).

The considerations on languages to be found in this book are very general and, e.g., do not depend on the notation used in the recording of expressions in the languages under consideration.

The influence of Polish logic is reflected in the strictly formal manner of presentation, and in particular in the use of the axiomatic method. It is commonly known that those important factors dominated the enquiries of eminent representatives of the Warsaw school of logic, such as Stanisław Leśniewski, Jan

Łukasiewicz, Alfred Tarski, Jerzy Słupecki, and others.

Reference to Polish semiotics manifests itself in the *categorial approach* to language, i.e., the approach which in the syntactic description of language takes into account the general principles of Leśniewski's theory of syntactic categories [1929,1930] in the version improved by Kazimierz Ajdukiewicz [1935].

It is worth noting that Leśniewski's theory of syntactic categories was constructed under the influence of the ideas of pure grammar, advanced by Edmund Husserl [1900,1901], and also under the influence of the concept of logical type, for the language of protothetics and ontology [1929,1930]. That theory formulates general principles of the classification of expressions into syntactic categories, its main objective being the formulation of criteria of the meaningfulness,that is well-formedness of expressions. In that spirit it was developed or used by Tarski [1933], Ajdukiewicz [1935], Innocenty M. Bocheński [1947], Tadeusz Kubiński [1958], Witold Marciszewski [1977, 1978a], and others. For similar purposes, although independently of Leśniewski, a theory of syntactic categories was presented and developed for the needs of the so-called combinatory logics by Haskell B. Curry [1961,1963].

The categorial approach to language, used in the present book, refers to Ajdukiewicz's conception [1935] and also to his ideas to be found in [1960], and is based on the method of indexation of expressions, that is the method of assigning them categorial indices and the resulting defining of their syntactic categories. That method makes it possible to apply the algorithm which allows one to examine the well-formedness, i.e., syntactic correctness, of expressions (in Ajdukiewicz's terminology, their syntactic connectedness). The categorial approach makes it possible to treat language as a language generated by the appropriate *categorial grammar*, whose idea goes back to Ajdukiewicz [1935].

The unquestionable achievements of Yehoshua Bar-Hillel [1950, 1953,1964], who shaped the concept of categorial grammar and popularized Ajdukiewicz's conceptions, and thus made a great contribution to the development of the foundations of categorial grammars, and the equally unquestionable attainments in that respect of Joachim Lambek [1958,1961] and the continuators of these two prominent researchers, bear a certain relation to the approach of the foundations of categorial languages suggested in this book, even though they do no play any essential role.

The philosophical trend of the theoretical reflections to be found in the present book manifests itself in the biaspectual treatment of language: as *language of expression-tokens* and as *language of expression-types*, the latter being interpreted in an abstract way. This is accompanied by the claim that it is the expression-tokens, accessible to sensory cognition, which are the basis of considerations on language, while the abstract expressions (expression-types), of which the former are representations, are derivative and definable constructs. This approach is linked to the *nominalistic (concretistic)* standpoint in the philosophy of language, represented in particular by such Polish logicians and philosophers as S.Leśniewski, T.Kotarbiński, and also J.Słupecki (during the last years of his life).

These aspects which determine the main assumptions made in this book, namely the logical, the linguistic, and the philosophical aspects, result in a two-level formalization of both theories of categorial languages, or theories of categorial grammars. Formalization at the first level pertains to languages of expression-tokens and yields the theory **TSCL** of simple syntactically categorial languages (Chap.II) and the theory **TSCω-L** of the ω-languages of such expressions (Chap. III). Formalization at the second level pertains to expression-types and yields, respectively, the dual theories **DTSCL** and **DTSCω-L** (Chap.IV). Those theories are expansions of the relevant theories formalized at the first level.

One of the basic objectives is to demonstrate the mutual relationships between theories formalized at these two different levels.

1.2. Comprehension of the formalisms which yield the said theories requires only an elementary knowledge of mathematical logic in the broad sense of the term, but it assumes a certain skill in the consistent application of the formal logical apparatus. This is why, in order to make the Reader better acquainted with the subject matter of the book we shall formulate below (in Sec.2) the intuitions pertaining to the philosophical and linguistic aspects of the framework used and to the formalisms themselves. They may prove useful in explaining the main ideas underlying the two suggested formal theories of the syntax of language. The understanding of the book should also be facilitated by numerous verbal comments, referring to intuition, in most cases accompanying the symbolic recordings of axioms, definitions, and theorems (the proofs of the latter are, in the overwhelming majority, transferred to the Annex to make the perception of the main text easier).

1.3. The thus outlined two-level *system* of treating both theories of language has, as it were, its precedent in the theory of algorithms formulated by A.A.Markov [1954], which, however, is not an axiomatic theory. An endeavour to axiomatize it was undertaken by Jerzy Słupecki. It was on his initiative and under his guidance, and with reference to Markov's ideas, that the axiomatic theory of concrete and abstract words was formulated (see G.Bryll and S.Miklos [1977]). It was followed - for other purposes, connected with the theory of syntactic categories - by the axiomatic theory **TLTk** of label tokens, constructed by the present author (see [1981a,1983a]). These goals coincided in the 1970's with the revived interests in categorial grammars, which had their semantic and philosophical grounds. The fact was stressed by Witold Marciszewski [1978b] and found manifestations in publications of such authors as P.Geach [1970], M.J.Cresswell [1972,1976], D.Lewis [1972], R.Montague [1970,1973], and A.Nowaczyk [1978]. That trend in research had been anticipated by R.Suszko [1958,1960].

The theory **TLTk** has become the basis for considerations on syntactically categorial languages. It also forms the core of all the theories of language presented in this book and mentioned previously. It is shown in Chap.I. It is also an outline of the intuitions connected with its formalization that opens the preliminary presentation of the contents of the book (Subsection 2.2).

1.4. But first a brief explanation is needed of the relation which the English-language version bears to the Polish original. The essential content of the book has remained unchanged. Only the Introduction, the References and the Final Remarks (which now form a separate item that follows Chap.IV) have been expanded. This has been done for three essential reasons: to make the book accessible to a much broader category of Readers, to bring it closer to the world literature of the subject, and to outline new solutions thematically connected with the original version of the book. This, it seems, places the book in a better perspective, especially in the light of the recent achievements in the field of categorial grammars, whose turbulent development originated in the 1950's and, revived in the 1970's, has continued up till now. It must be added here that this development has been influenced by the still vital research on formal grammars in the spirit of Noam Chomsky, going on incessantly since the publication of his book [1957] that marked a breakthrough in linguistics.

It would not be possible here to indicate, with a view to referring the present book to historical and present studies in the sphere of categorial grammars, the differences and the coincidences between this book and the said studies. This is why I take the liberty to refer the Readers above all to the entry "Categorial grammar" in the *Dictionary of Logic* edited by Marciszewski [1981], the anthology bearing the same title and edited by Buszkowski, van Benthem, and Marciszewski [1988], and the selected publications by the leading contemporary students of categorial grammars, W.Buszkowski [1982a-c,1984,1985,1986] and J.van Benthem [1984,1986], which appeared when the Polish version was being prepared for the press or after its publication.

The essential part of the original version was ready in 1981, and the main ideas of the underlying theoretical conceptions were presented at the Conference on the History of Logic (see [1981b]), held in Cracow and dedicated to the scientific production of Kazimierz Ajdukiewicz. The Polish version appeared in 1985. Professor Jerzy Słupecki, who suggested to me the problems analysed in it, had died before the publication of the English-language version. Let me, therefore, dedicate it to his memory.

2. Intuitive foundations

2.1. The controversy over the problem of the double ontological natúre of linguistic objects is linked to the well known controversy over universals. The words written in the preceding sentence are *concrete* physical objects perceivable by sight. The second and the twentieth word which occur in the first sentence in this paragraph, like the word occurring within the quotation mark in the term "controversy" may be treated as concrete *equiform* representations of that name. That name is often treated as the name of a certain abstraction,a *word type*, that is the class of all the words which are equiform with the word in the quotation marks. The class of such concrete words is then an abstract word, or a certain ideal entity. The controversy over the ontological nature of the objects designated by the name "controversy" arises when we want to decide which of the kinds of words denoted by that name has original existence, and which has derivative existence.

Without making any decision in that matter we shall point to the fact that the demonstrated duality in the interpretation of words or expressions of language, which has become common especially owing to the important study of A.Tarski [1933], takes place in logic and semiotics. Note also that following the distinction between *token* and *type* made by C.S.Peirce [1931-1935] (see also Y.Bar-Hillel [1953,1954]) inscriptions, words, expressions understood as concrete objects are included in the class of *tokens*, and when understood as classes of equiform tokens - are included in the class of *types* (R.Carnap [1942] used, respectively, the terms *events* and *designs*).

In accordance with the well known division made by Ch.Morris [1938] logical semantic analysis can be carried out in three aspects: the syntactic, the semantic, and the pragmatic. Such an analysis is sometimes applied to concrete and sometimes to abstract linguistic objects. For instance,in syntactic studies, when formulating rules of constructing coherent expressions we make use of expression types, but if we want to examine the grammatical correctness, i.e., syntactic connectedness, of expressions and use for that purpose, e.g., Ajdukiewicz's algorithm [1935] we make use of tokens. At the pragmatic level linguistic analysis pertaining to the functioning of language is linked to the use of expressions in definite contexts and situations (see J.Pelc [1967], [1971]). Reference to the numerous functions of linguistic expressions (descriptivity, emotiveness, suggestiveness, decisiveness), substantiation of statements, and in particular theorem proving requires the use of expression tokens. On the contrary, the functioning in pragmatics of the concept of communication is based on the concept of expression type (potential inscription). It is the same when it comes to such basic semantic concepts as designation, denotation, truth, and meaning.

The said ambiguity in the use of linguistic inscriptions in the analysis of language has also its broader dimension. This is so because in the literature of the subject it is not always clear whether expression type is understood as a class of equiform tokens, sometimes there is even no reference to the concept of equiformity even though the context indicates that expression types are meant. W.Marciszewski claims that the concept of *type* has several different interpretations and need not be understood in an abstract way.

Note in this connection that when speaking about the dual ontological nature of linguistic objects we shall refer to *tokens* and *types* in accordance with the distinction made by Peirce.

The dual ontological nature of linguistic inscriptions and the possibility of using them in two ways in semiotic analysis results in the necessity of the biaspectual description of language in the theoretical, logical conception of language, as a language of expression tokens and a language of expression types. Yet in formalizing a given conception of language we must choose either tokens or types as the point of departure.

In the two approaches to the theory of language suggested here we refer to the *nominalistic (concretistic)* interpretation and construct those theories first as the theories **TSCL** and **TSCω-L** of languages of expression tokens (the first level); in doing so we start from token (*label token*) as a primitive, undefined, concept and several other concepts connected with the former. When constructing those theories as theories of expression types (the second level) we obtain the dual theories **DTSCL** and **DTSCω-L**, respectively, in which the fundamental concepts include such abstract concepts as label type, word type, expression type, which are defined and derived constructs.

2.2. The theory **TLTk** of label tokens (Chap.I) is the point of departure for the formalization of both conceptions of the syntax of language. In that theory four concepts appear as primitive, undefined, whose meaning is explained by the appropriate axioms. They are: the *set Lb of all label tokens*, the *relation* $\approx$ *of equiformity*, the *vocabulary V*, distinguish from the set *Lb*, and the *relation* c *of concatenation.*

The set *Lb* is the universe of an arbitrary but fixed simple language $\mathscr{L}$ or ω-language ω-$\mathscr{L}$, which has tokens as its expressions. The elements of *Lb*, called *label tokens*, are concretes and may (but need not) be inscriptions. Label tokens are any objects perceivable by sight, extensive in time and space, from which by *equiformity* one can obtain elements of expressions of graphic languages. Out of label tokens we choose the simplest words of the language $\mathscr{L}$ or ω-$\mathscr{L}$, namely the words which form the *vocabulary V* of $\mathscr{L}$ or ω-$\mathscr{L}$, and from them in turn we form compound words of that language, the expressions, including the well-formed expressions of which consist $\mathscr{L}$ or ω-$\mathscr{L}$. Label tokens also include the categorial indices which make it possible to determine the syntactic categories of expressions, to demonstrate their syntactic roles, and to examine grammatical coherence. The existence of label tokens and their varietes listed above follows from the axioms adopted in the theories **TSCL** and **TSCω-L**.

The *relation of equiformity* holds between pairs of label tokens. It is interpreted very broadly by imparting it a pragmatic sense: it depends on the objective which we have in view in order to obtain definite results. The equiformity of label tokens may, but need not, be determined by physical similarity. For instance, for a visual artist, a graphologist, or a printer the concrete words: ontology, *ontology*, ONTOLOGY (lower case, lower case italics, capitals) are diversiform, whereas from the point of view of a person who is learning the meanings of words they may be treated as equiform. On the other hand, in categorial analysis intended to examine the syntactic correctness of expressions, for instance of the tautological sentence

The creative man is creative

the adjectives which occur in it may be treated as diversiform in view of their enviroment: the first modifies a noun and is a name-forming functor, while the second is independent and, being a name, belongs to another syntactic category and has a different categorial index (see H.Hiż [1960,1967,1968]). Should we resort to a metaphor, we might say that it is like with one and the same deciduous tree in summer which differs from itself in winter when it has no leaves. Note also that a word which belongs to the vocabulary of a natural language or even a formalized language analysed at the *token level* is equiform with at least one word used earlier in the process of communication and has a categorial index equiform with the categorial index of the former, and hence also the same grammatical category.

It is further assumed that the relation $\approx$ of equiformity is reflexive, symmetric, and transitive, without considering the problem whether that postulate could be replaced by a less rigorous one. It is also postulated that a label token which is equiform with a word from the vocabulary of a language $\mathcal{L}$ or ω-$\mathcal{L}$ is also a word from the vocabulary of that language.

The *vocabulary* V is used to generate the *set* W *of all the words* of $\mathcal{L}$ or ω-$\mathcal{L}$. That language is determined by the *set* S *of all its meaningful*, i.e., *well-formed expressions*, which is a subset of W. The set W is generated from the vocabulary V by the *relation* c *of concatenation*, which holds between any two label tokens and the label token which is their *concatenation*. W is the least set which includes V and every label token which is the concatenation of simple word tokens, i.e., the ordinary concatenation of two simple words from the vocabulary V, or the *generalized concatenation* of more than two simple words from V.

At the *token level* the relation of concatenation need be a function because two given label tokens may yield many equiform concatenations. This is so because intuitively concatenation may be understood as the relation of writing, on the right side (e.g., in European ethnic languages) or on the left side (e.g., in Semitic languages), and at the same level, of a label token equiform with the first label token in a given pair, of a label token equiform with the second label token in that pair. But it may also be a relation of non-linear juxtaposition of two label tokens which thus form a single label token. This may take place, e.g., in hieroglyphic script and in many exact sciences. Upon the relation of concatenation we impose axiomatically certain intuitive conditions related to A.Tarski's axioms of metascience [1933] (see also e.g. H.Hiż [1957]), but in doing so we have to bear in mind the fact that in our case that relation is defined for tokens, and not for types. The axioms of the theory **TLTk** guarantee, of course, that the words from the vocabulary V of $\mathscr{L}$ or ω-$\mathscr{L}$, unlike the compound words from the set $W \setminus V$ of that language, are not concatenations of any two label tokens.

The *generalized concatenation* of label tokens is an ordinary concatenation or the concatenation of an ordinary concatenation and a given label token. A compound word of $\mathscr{L}$ or ω-$\mathscr{L}$ is thus always a generalized concatenation of simple word tokens, because it can be obtained by a single or multiple application of ordinary concatenation to simple words from V. A compound word is thus formed of a finite number of vocabulary words. The number of such words which are parts of a given words is defined as its length.

2.3. The second stage in the formalization of the theories **TSCL** and **TSCω-L** of categorial languages $\mathscr{L}$, and correspondingly ω-$\mathscr{L}$, consists in the construction of the theory **TETk** of expression tokens. Its presentation is to be found in Chap.II where the theory **TSCL** is shown. Now the theory **TETk** is obtained from **TLTk** by the introduction of two new essential concepts, that of *expression token* and that of *categorial index*. Here are the principal intuitions connected with the formalization of that theory.

Categorial indices are used to carry out linguistic categorial analysis in accordance with the ideas advanced by

Ajdukiewicz [1935,1960]. Such an analysis pertains exclusively to any *compound expression tokens* that form the subset E_c of the set W of word tokens. In accordance with the principle originating from G.Frege [1879] every expression from the set E_c has the functor-argument structure - one can distinguish in it its constituent called the main functor and its constituents called the arguments of that functor.

The proper categorial analysis of a compound expression token (the examination of its syntactic correctness) is preceded by its *functorial analysis*: after having distinguished in a given compound expression the main functor and its arguments we continue the process of distinguishing the main functor and its arguments with reference to all constituents of a given expression which are compound expressions. By way of example, the functorial analysis of the sentence "John ardently loves Helen" is illustrated by two diagrams of a tree. In the first, the bold lines indicate the main functors; in the second the method of parsing used by Ajdukiewicz is taken into consideration (see Fig.1 and Fig.1a below).

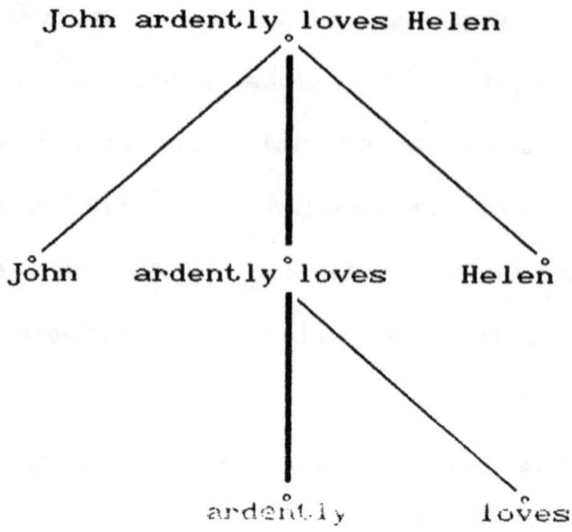

Fig.1

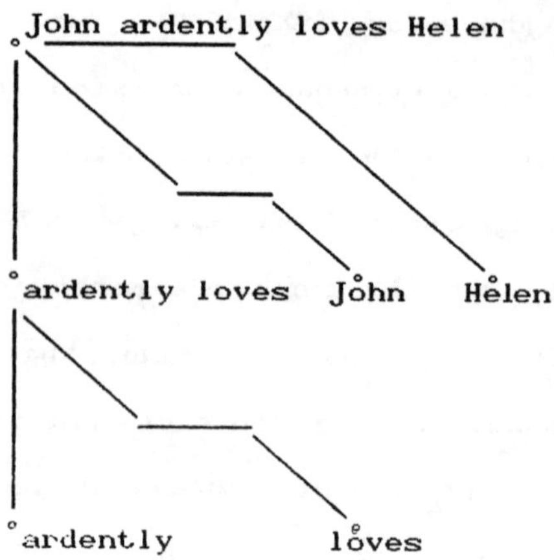

Fig.1a

The categorial indices (types) by means of which we examine the syntactical correctness of compound expression tokens are label tokens of *Lb*. But they are not in the set *W* of the words of $\mathscr{L}$ or ω-$\mathscr{L}$. They are words from metalanguages of those languages and, to use Buszkowski's terminology, are used for the *typization*, or determination of the syntactic categories, of expressions in those languages and for the examination of their grammatical correctness. For the formal syntactic description of the categorial language $\mathscr{L}$, or ω-$\mathscr{L}$, we introduce a new primitive concept, namely that of *auxiliary vocabulary* I_o that is the *set of all basic indices*, which includes at least one technical symbol, e.g., slant /, colon :, comma ,, bracket (or), etc. It is, of course, postulated, that the sets I_o and *V* have no common elements. From the vocabulary I_o we generate, by the relation c of concatenation, the *set I of all categorial indices*, as the set W is generated from V. The set I thus contains all basic indices from I_0 and all indices of functors. The latter are generalized concatenations of basic indices. The principles of the construction of the indices of functors of a given categorial language $\mathscr{L}$ or ω-$\mathscr{L}$ are fixed by special rules. For instance, of the basic indices s and n, used, respectively, for the typization of sentences and names in such a language, and the auxiliary symbol /, on the assumption that the concatenation of label tokens provides for right-side linear juxtaposition, we can form indices of functors of a quasi-fractional form, e.g., s/nn and s/nn//s/nn, which are, respectively the index of the sentence-forming functor of two name arguments, and the functor-forming functor, which has the above functor as its only argument and forms such a functor.

The typization of definite words of $\mathscr{L}$ or ω-$\mathscr{L}$ is obtained by the *relation* ι *of indication of indices of words*, in other words, the *relation of typization*. That relation is a new primitive concept in the theory **TETk**. It is postulated that it is a function which assigns to definite word tokens one index token. It is also axiomatically assumed that words which are equiform with a word that has an index have their indices equiform with that of the said word.

From the set of those words which have indices, that is from the domain $D(\iota)$ of the relation ι, we distinguish the *set* E *of all expression tokens* of $\mathscr{L}$ or ω-$\mathscr{L}$, and hence both the *set* E_s *of all simple expressions* of such a language, which are words of V and have an index, and the *set* E_c (obtained from E_s) *of all compound expression tokens* of that language. The principles of the formation, out of words that have indices, of compound expressions of such a language are fixed by its syntactic rules. In theoretical considerations those rules are replaced by a single one, namely the one-to-one *function* ρ *of the formation of compound expression tokens* of $\mathscr{L}$ or ω-$\mathscr{L}$. The set E_c is defined as the counter-domain $D^{-1}(\rho)$ of that relation.

The function ρ is the third and the last primitive concept of the theory **TETk**. We associate with it the following intuitions. It allows us to assign to any finite configuration $\langle p_0, p_1, \ldots, p_n \rangle$, where $n \geq 1$, of words of $\mathscr{L}$ or ω-$\mathscr{L}$ that have categorial indices, such a generalized concatenation of them which in the language of **TETk** is recorded as

$$(\delta) \qquad \rho(p_0, p_1, \ldots, p_n)$$

and termed a compound expression token of $\mathscr{L}$ or ω-$\mathscr{L}$ consisting of the words $p_0, p_1, \ldots, p_n$. This is so because (δ) is treated as a substitute of any compound expression of such a language which is constructed of the main functor p_0 and its successive arguments $p_1, \ldots, p_n$. In doing so we do not consider the formal structure of such an expression and disregard the symbolism in which it is recorded and how it is formed as the appropriate generalized concatenation by rules valid in $\mathscr{L}$ or ω-$\mathscr{L}$. The inscription (δ) thus may denote various expressions of $\mathscr{L}$ or ω-$\mathscr{L}$, built according to various rules, on the condition that in each such expression the number of words is the same and that among those words we can distinguish the main functor and its arguments. Moreover, such expressions need not be of one and the same syntactic category; they may be sentences, names, and functors. Further, (6) may also replace different but synonymous expressions recorded in different notations (see Sec.II.4).

The theory **TETk** is enriched by new concepts, specific of **TSCL** and **TSCω-L**, the most important of which is that of a *meaningful expression*, i.e., *well-formed expression token*. That concept is defined differently for $\mathscr{L}$ and for ω-$\mathscr{L}$.

2.4. The theory **TSCL** describes any simple categorial language $\mathscr{L}$ of expression tokens. Its *well-formed expressions* are elements of the set S, which in that theory is defined as containing all simple expressions from E_s and all those compound expression tokens of $\mathscr{L}$, which have the property that both they and each constituent of theirs which is a compound expression of that language satisfy the *rule* $\mathfrak{m}$ which in a free formulation states:

The categorial index of the main functor of a compound expression is obtained as a generalized concatenation from the index of the expression, i.e., the index of the expression which that functor forms together with its arguments, and the indices of all successive arguments of that functor.

This definition of well-formed expression makes it possible to describe an algorithm of examining the syntactic correctness of linguistic expressions which comes close to Ajdukiewicz's algorithm [1935]. This is so because categorial analysis consists in finding whether the rule $\mathfrak{m}$ holds for every constituent (which is an expression of $\mathcal{L}$) of a given expression of $\mathcal{L}$ that consists of functors and arguments. If this is so, then a given compound expression of $\mathcal{L}$ is in S and hence - is a well-formed expression. By referring to the above example of an expression whose functorial analysis was illustrated by Fig.1 and Fig.1a, we can easily check its well-formedness by assigning categorial indices to its constituents. For that purpose we can use the tree of categorial indices of that expression shown in Fig.2 and Fig.2a below, where s and n, and the quasi-fractional inscriptions are interpreted as explained previously.

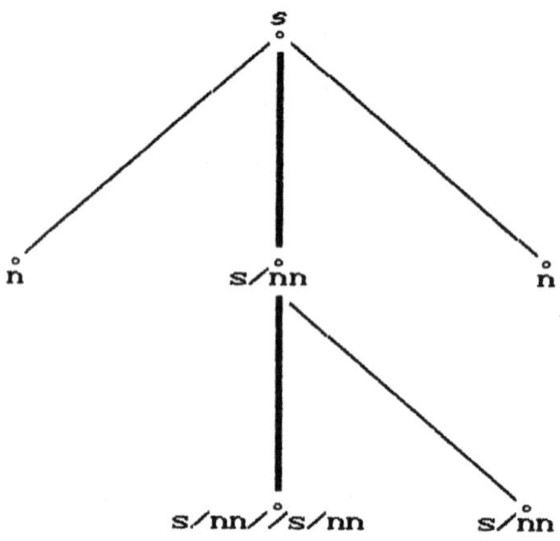

Fig.2

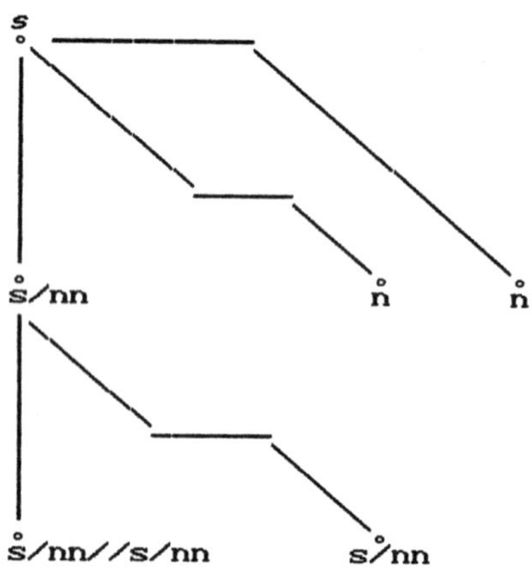

Fig.2a

It can also easily be seen that the set S of all well-formed expressions of a categorial language $\mathcal{L}$ can be generated in this way by the system

$$G_{\mathcal{L}} = \langle E_s, I_o, \iota, \mathfrak{m} \rangle,$$

which may be treated as a reconstruction of *classical grammar*, whose idea goes back to Ajdukiewicz [1935]. Such a grammar is *rigid* (see W.Buszkowski [1986]), which is to say that a given expression of $\mathcal{L}$ has only one index (type), and hence only one syntactic category, assigned to it. It will be assumed about a syntactically described categorial language $\mathcal{L}$ that it is *unambiguously typizable* (see W.Buszkowski [1986]). Such an assumption is quite natural because finitely typizable languages, which we usually consider to be natural languages, can always be disambiguated by the "separation" of words or expressions to which a finite number of indices (types) is assigned. That can be achieved if we provide them, for instance, with the appropriate numerical indicators, letters, or diacritical signs. Moreover, as has been said earlier, we can very broadly interpret the concept of equiformity, which substantiates the assumption made above.

The categorial nature of the language $\mathcal{L}$ is also connected with the logical partition $Ct(S)$ of the set S of all well-formed expressions of $\mathcal{L}$ into non-empty and pairwise disjoint *syntactic categories*. Syntactic categories are determined by categorial indices and are sets of expressions which have equiform categorial indices. In the family $Ct(S)$ of all syntactic categories we can distinguish *basic categories* and *functoral categories*. The former include all *basic expressions*, i.e., expressions with basic indices from I_o, while the latter include *functors*, i.e., expressions with functoral indices from the set $I \setminus I_0$. The set of all basic expressions of $\mathcal{L}$ and the set of all its functors have no common elements and the sum of those two sets cover all well-formed expressions of $\mathcal{L}$.

The concept of syntactic category bears a certain relation to the relations of *replaceability* of expressions, which is a defined concept in **TSCL**. In the traditional interpretations a syntactic category is the set of all expressions mutually replaceable in arbitrary sentential contexts, or, more generally, in all well-formed contexts. The traditional definitions of that concept (see S.Leśniewski [1929], A.Tarski [1933] , K.Ajdukiewicz [1935] , I.M.Bocheński [1949]) imply problems because they do not preclude a vicious circle, which is due to the fact that the concepts of sentence or meaningful (well-formed) expression and that of replaceability of expressions function in them as undefined. Moreover, as was stated by R.Carnap [1937] the principle of replaceability of expressions in arbitrary well-formed contexts, assumed in the traditional definitions of syntactic categories, may be questioned (cf. also H.Hiż [1961]). The correctness of such definitions can be refuted by examples. In academic handbooks of logic, e.g., the definition of names guarantees to proper names and personal pronouns belonging to the category of names, whereas it can easily be seen that well-formed expressions, respectively, a nominal and a sentential one (see J.Lambek [1958]),

Big John, Big John is here,

on the replacement in them of the proper name by the personal pronoun "he" yield meaningless (not well-formed) expressions

Big he, Big he is here.

On the other hand, for instance, both expressions "2=2" and "2+2" are well-formed, but the latter is obtained from the former by the replacement in it of the sentence-forming functor "=" by a functor belonging to another syntactic category, namely the name-forming functor "+".

Even though we deviate here from the traditional definitions of the concept under consideration it remains in an important connection with *replaceability*. That connection is described in Sec.II.7 by the *fundamental theorems of the theory of syntactic categories*, at first intuitively discussed in Sec.II.1, where the objectives of the construction of the theory of categorial languages are outlined.

2.5. The theory TSCω-L describes an arbitrary but fixed categorial ω-language ω-$\mathcal{L}$ of expression tokens. Its well-formed expressions belong to the set S, defined in a slightly different way than in the theory TSCL, because its expressions may include *variables* and *operators* that bind them. The definition of S should imply all that as has been shown in 2.4, makes it possible to carry out the categorial analysis of the expressions of ω-$\mathcal{L}$ that could be correct from the point of view of the theory of syntactic category. Such a definition should accordingly enable us, in particular, to formulate an algorithm of the examination of the syntactic correctness of expressions containing operators and variables bound by them. The endeavours to formulate such an algorithm, undertaken already by Ajdukiewicz, as far as I know deviate more or less from the general principles of the theory of syntactic categories. An algorithm which refers to those principles is described in Sec.III.6.

In order to give a proper definition of a well-formed expression in ω-$\mathcal{L}$ we add to the theory **TETk** new concepts, which do not occur in **TSCL**. They are those of the *set O of all operators*, the *set Vr of all variables*, and the *relation (fv) of being a free variable* in a given expression token. The first two of them are ptimitive concepts in **TSCω-L**. The intuitive sense which we impart to *operators* (elements of the set *O*) and to *variables* (elements of the set *Vr*) does not deviate from the interpretation of these concepts in the exact sciences. To simplify the matters we assume that one operator binds only one variable while its scope is limited to only one expression token. Operators and variables are treated as simple expressions in ω-$\mathcal{L}$. Operators as such are elements of the vocabulary of ω-$\mathcal{L}$ and cannot be concatenations of any label tokens, and moreover they can be assigned only one categorial index each. Thus, for instance, we treat as an operator the existential quantifier $\vee$ or $\exists$, and not the graphic symbol $\bigvee_{x}$ or $\exists x$, and we can assign it one categorial index (the same applies to any inscription equiform with it). This seems to be in contradiction with the so-called typical ambiguity of operators (see, e.g., R.Suszko [1958], [1964]), with their being treated as categorial open functors (see D.R.Vanderveken [1976]), and as context-dependent functors (see, e.g., W.Marciszewski [1989]). For such an interpretation, graphically similar quantifiers occurring, for instance, in the following two arithmetical expressions and an expression of the sentential calculus with quantifiers:

(i) $\bigvee_{x} x \leq y$, $\bigvee_{x} \bigwedge_{y} x \leq y$, $\bigvee_{p} (p \rightarrow \neg\neg p)$

have different syntactic categories and hence also different categorial indices. The first two because in Marciszewski's [1989] interpretation the propositional functions $x \leq y$ and $\bigwedge_{y} x \leq y$ have different syntactic categories, and the third because, unlike the first two, it binds the variable p which has the category of a sentence, and not of a name. This problem can be eliminated as previously (see 2.2), either by treating those quantifiers as diversiform because of their environment, or by providing them with some diacritical signs which would make it possible to treat them as diversiform.

Problems with assigning to the operators that bind variables their syntactic categories have been known since the times of Leśniewski [1929], Tarski [1933], and Ajdukiewicz [1935]. In the interpretation suggested here they will be treated as functors of a special kind, which according to the context have a definite categorial index each. That binding role of theirs manifests itself in definite expressions of ω-$\mathscr{L}$ called *operator expressions*; they are ordinary compound expression tokens consisting of three constituents each: an operator, called the *main operator* of that expression, a variable called *indexical variable*, and an expression token called the *scope of that operator*. Thus the main operator is the main functor of such an expression, while the indexical variable and the scope of the operator are its successive arguments.

The operator in an operator expression may be *empty-binding* if in its scope it has no *free variable* or if it *does not bind* any *free variable* in its scope (which is to say that no free variable in its scope is equiform with its indexical variable). The concept of free variable in a given expression is, of course, defined by means of the relation (*fv*), namely as a variable which bears the relation (*fv*) to that expression, if it is a constituent of that expression and does not occur at the same time in the scope of any operator in which it would be equiform with its indexical variable.

The definition of the *set S of all well-formed expression tokens* of ω-$\mathcal{L}$ takes into consideration the degree of the complexity of the expressions which are its elements. The simplest *well-formed expressions* in that set are all its *simple expressions* (including operators and variables). The simplest compound *well-formed expression* in that set is either 1° a *non--operator expression* which is a compound expression token of ω-$\mathcal{L}$ consisting of simple expressions and such that its main functor is not an operator and also such that it satisfies the rule $\mathfrak{m}$, or 2° an *operator expression* such that the scope of its main operator is the simplest non-operator expression containing the variable which that operator binds, and such that this expression also satisfies the rule $\mathfrak{m}$ (the index of the main operator of that expression is a concatenation of the index of that expression and the indices of the indexical variable of the main operator and that of its scope).

More complex *well-formed non-operator expressions* differ from the simpler well-formed non-operator expressions only by being constructed of at least one compound (operator or non--operator) well-formed expression, and more complex *well-formed operator expressions* differ from simpler ones by the fact that their scope is not a simplest well-formed compound expression. Note that the set S thus includes all simple expressions of ω-$\mathcal{L}$, all its non-operator expressions, and all its operator expressions. Note also that every compound well-formed expression of ω-$\mathcal{L}$ and every constituent of such an expression satisfy the rule $\mathfrak{m}$, which makes it possible to generate the set S of ω-$\mathcal{L}$ by the categorial grammar

$$G_{\omega\text{-}\mathcal{L}} = \langle E_{\aleph}, I_{\circ}, \iota, \mathfrak{m} \rangle,$$

and also to describe the algorithm of the examination of the syntactic connectedness of expression that resembles Ajdukiewicz's algorithm for expressions without operators that bind variables.

When applying that algorithm we start from the parsing of a given compound expression token of ω-$\mathcal{L}$ by carrying out the functorial analysis of that expression as described under 2.3. If functorial analysis brings us to simple expressions (which is to say, if the ends of the tree are simple expressions) of

ω-$\mathscr{L}$, then we find whether there is among them an empty-binding operator. If it is not so, then we apply categorial analysis in the proper sense of the term by assigning to every expression obtained by parsing its categorial index and check whether every compound expression thus obtained satisfies the rule $\mathfrak{m}$. If this is so, then the given compound expression of ω-$\mathscr{L}$ is well-formed. A precise description of that algorithm and examples of its application are to be found in Sec.III.6.

Note also that if we apply the same designation with the index s of sentences and propositional functions and if we assign the same index n to names and name functions (and use the same principles of forming index functors as in 2.3 and 2.4), then the quantifiers which occur in the first two expressions in (i) have the index s/ns, whereas the quantifier which occurs in the third expression in (i) has the index s/ss. This is not at variance with the fact that the syntactic category with the index s/ns covers such cases in indirect speech as "thinks that" and "knows that", while the category s/ss covers truth functors of two arguments each. This is so because we do not adopt here the traditional definition of syntactic category. By the way, in order to distinguish the syntactic category of closed sentences from that of propositional functions in which one, two, etc., free variable occur, we can accordingly use the typization: s, s_1, s_2, etc. In such a case the said existential quantifiers would have the indices s/ns_1, s/ns_1 and s/s_1s_1, respectively, and the universal quantifier in the second expression in (i) would have the index s/ns_2 (see W.Marciszewski [1989]). The above outlined algorithm obviously does work in such a case.

It must be emphasized at this point that all the properties which make it possible to describe, in terms of **TSCL**, language $\mathscr{L}$ as a categorial language are applicable to language ω-$\mathscr{L}$ as well. In the theory **TSCω-L** all the analogues of the theorems for well-formed expressions of $\mathscr{L}$ hold, too.

2.6. The formalization of the dual theories **DTSCL** and **DTSCω-L** takes place at the second level, at which we are interested in types and relations among them. Those formalization pertain, respectively, to an arbitrary but fixed simple language $\overline{\mathscr{L}}$ of expression types and an arbitrary but fixed ω-language ω-$\overline{\mathscr{L}}$ of such expressions.

All the linguistic elements of expression tokens of $\mathscr{L}$ or ω-$\mathscr{L}$, e.g., label token, word token, expression token, well-formed expression token, variable token, operator token, and object token which make it possible to carry out the categorial analysis of expression tokens, and hence categorial index tokens, have analogues in induced concepts which are elements of expression types in $\overline{\mathscr{L}}$ or ω-$\overline{\mathscr{L}}$, namely label type, word type, expression type, etc. and, for categorial analysis, index types. All of them are defined as classes of abstraction of the corresponding equiform tokens.

Likewise, all the relations holding among tokens have their analogues in induced relations holding among types. The latter are defined so that they hold among types if and only if relations with the same intuitive meaning hold among their representations at the token level (representatives of classes of abstraction).

All the properties of concepts grasped by the axioms, definitions, and theorems of **TSCL** (or **TSCω-L**) which describe categorially the syntax of $\mathscr{L}$ (or ω-$\mathscr{L}$) of expression tokens pass upon the induced concepts. This makes it possible to describe categorially the syntax of $\overline{\mathscr{L}}$ (or ω-$\overline{\mathscr{L}}$) of expression types. We find the full analogy between the syntactic concepts of the languages in question analysed at two different levels. Since in the syntactic description of language the assumption of the existence of object tokens as the fundamental linguistic elements served as the point of departure, we can conclude from the above that a theoretical categorial syntactic description of language does not require the preliminary assumption that language is a language of abstract expressions, that is a language of certain ideal entities. Hence in syntactic analyses of categorial language we may abstain from making assumptions about the existence of ideal linguistic objects.

This conclusion can be substantiated more strongly in the light of recent studies carried out by the present author (see Final Remarks).

Acknowledgements

Professor Grzegorz Bryll was the first reader of the typescript of the present book in its original version and checked the formal and factual correctness of my work. I owe him my cordial thanks for that.

My thanks are also due to Professor Jan Srzednicki for his initiative to have the book published in English. I am particularly grateful to Professor Stanisław J. Surma for his efforts accompanying the publication of the English version. I also thank all those nameless friends who have contributed to its appearance.

CHAPTER I

THE AXIOMATIC THEORY TLTk OF LABEL TOKENS

Sec. I.1. Primitive concepts

Let us imagine the situation of a user of an ethnic language who wants to write a letter or the situation of a mathematician who wants to formulate theorems or to produce proofs in a formalized language. Both of them select and write out simple labels of a definite shape, combine them into more complex labels that follow one another, and frequently reproduce them. Linguistic expressions used by a given language user are concrete, physical, sensorily perceivable objects, namely *label tokens* (in Latin (labels) *in concreto*, in English also label events; see Ch.S.Peirce [1931-1935]; R.Carnap [1942]). A precise explanation of the properties of those material objects encounters greater difficulties than those which we face when we establish the properties of ideal, abstract objects. We face such difficulties when we want to formulate an intuitive characteristic of a given language.[1]

How are we to characterize syntactically any language $\mathscr{L}$? We usually begin with writing out or indicating simple standard labels. Thus our point of departure in characterizing $\mathscr{L}$ consists in fixing the *vocabulary V*, which is a subset of the universal *set Lb of all label tokens* and consists of such simple standard words and those labels which are *equiform* with them. The set *S of all well-formed expressions* of $\mathcal{L}$ can now be defined as a certain subset, specified by the rules of syntax, of the *set W of all word tokens*, where W is the set of label tokens formed of the words from the vocabulary, and of labels obtained from the former by their *concatenation*. Thus, in a simplified formulation, the syntactic characteristic of $\mathcal{L}$ takes on the form of the ordered sixtuple

$$(\mathcal{L}) \qquad \langle Lb;\ V,\ W;\ \approx,\ c;\ S \rangle,$$

where $\approx$ stands for the relation of equiformity of labels, and c, for the relation of concatenation of labels.

The properties of the first five concepts of $(\mathcal{L})$ are fixed by the theory of label tokens to be presented in this chapter. Like in the earlier works by the present author [1981a,1983a] it will be symbolized as **TLTk**. The sixth concept, that of the set S of meaningful, i.e., well-formed expressions for the so-called *simple categorial language* and for the syntactically categorial language called *ω-language*, is described by the appropriate theories presented in Chap.II and Chap.III.

The problem of constructing an axiomatic theory of label tokens was formulated by Jerzy Słupecki and first solved by Grzegorz Bryll and Stanisław Miklos [1977] with reference to the theory of algorithms of A.A.Markov [1954]. The theory **TLTk** is thematically close to that formulated by Bryll and Miklos. It differs from the former primarily in those respects which are connected with its application in the theory of syntactic categories. It is thus a different solution of the problem formulated by Słupecki and is the foundation of the analyses presented in this book.

The theory **TLTk** is superstructured over the classical narrower functional calculus with identity and over set theory. It is based (cf. G.Bryll, S.Miklos [1977] and [1981a,1983a] by the present author; see also Sec.I.8) on four primitive concepts: those of sets *Lb* and *V* and those of the relations ≈ and c. The intuitive meaning of those concepts shall be discussed in greater detail in the successive sections; we shall also specify the conditions (axioms) which those concepts should satisfy.

Sec. I.2. Label tokens and their equiformity

Label tokens are empirical products. They are either material products of human beings or products of nature indicated by them. They are thus not only definite expressions of a given ethnic, formal, or formalized language, printed in a given place in a given copy of a book, or written down with chalk on a blackboard of with a fountain pen, ballpoint pen or pencil in a definite place on paper; they further include not only definite drawings, smoke and light signals, etc., but also contours of mountain ranges, frontiers of states, and configurations of leaves. The essential property of label tokens is that every part of a label token is also such a label.[(2)] The *set Lb of all label tokens* is a universal set, within which we single out certain sets of labels which make it possible for us to define a given language.

The *relation of equiformity* - another primitive concept of TLTk - is a binary relation in which *Lb* is both the domain and the counter-domain. Any two labels between which that relation holds are called *equiform.* Two equiform labels may occur in two different places. For instance,the thirteenth word in the first sentence and the seventh word in the second sentence in this paragraph are equiform labels. Thus the relation of equiformity is not an identity relation. Andrzej Grzegorczyk ([1981], p.112), when explaining the intuitive sense of the concepts needed for the description of various syntactic relationships, wrote in connection with this that "...An inscription *X* is *equiform* with an inscription *Y* if, when speaking in ordinary language, one would say that *X* and *Y* are two tokens of one and the same inscription. For instance, *X* has the form of the letter "A" and *Y* has the form of the letter "A", but one of those inscriptions is situated in a different place from that in which the other is ..."

The equiformity of labels is not always determined by geometrical similarity. Two words printed in different types but recorded with equiform letters of alphabet occurring in the same order may be equiform. The equiformity of expressions in a given language is related to the system of caligraphy.

The fact that the labels p and q are equiform (that is to say the relation of equiformity holds between them) is recorded as the expression $p \approx q$, which is read: the label tokens p and q are equiform, or: the label token p is equiform with the label token q.

An important property of the relation of equiformity is stated by the axioms:

AXIOM I.1

a. $p \approx p$,

b. $p \approx q \Rightarrow q \approx p$,

c. $p \approx q \wedge q \approx r \Rightarrow p \approx r$.

In the recording of these axioms we have availed ourselves of the Convention I.1

Convention I.1. The letters

(v_1) $p, q, r, s, t, u, v, \ldots,$

with or without subscripts, range over the set Lb.

The axioms I.1.a-c state that the relation of equiformity is an equivalence in Lb.

Sec. I.3. Concatenation

The *relation of concatenation* c is a ternary relation. All of its three domains equal Lb. A label which bears the relation of concatenation to two labels is called a *concatenation*. In the European ethnic languages, any concrete label obtained from a label equiform with p by writing on its right, on the same level, the successive label equiform with q is a concatenation of the labels p and q.

For instance, the label token

la Tour Eiffel,

which is a French language expression, and any equiform name obtained from the label tokens

Eiffel

la Tour

(the lower preceding the upper) is their concatenation. In Semitic ethnic languages concatenations of two label tokens are obtained in a similar way, but the second element is added on the left. Not in all languages concatenations are written down in a linear form. Numerous expressions of the language of the theory **TLTk**, which we shall use, are not recorded in that form (see, e.g., Axiom I.2). Note also that the relation of concatenation has the property that two equiform labels may be concatenations of the same labels, which shows that that relation need not be a function. These and other properties of the relation of concatenation are fixed by further specific axioms of **TLTk**. In recording them we shall use the expression $c(p,q,r)$, which is read thus: the label token r is a concatenation of a label token equiform with p and a label token equiform with q, or, more briefly, the label token r is a concatenation of label tokens p and q.

And here come the axioms which describe the relation of concatenation c:

AXIOM I.2

$$\bigvee_{r} c(p,q,r).$$

For any two label tokens there is a label which is their concatenation.

AXIOM I.3

$$c(p,q,r) \Rightarrow \neg (r \approx p) \wedge \neg (r \approx q).$$

A label token which is a concatenation of two label tokens is not equiform with any of them.

AXIOM I.4

$$c(p,q,t) \wedge c(r,s,u) \wedge p \approx r \wedge q \approx s \Rightarrow t \approx u.$$

Concatenations of two pairs of label tokens with first and second elements which are pairwise equiform are equiform labels.

AXIOM I.5

$$c(p,q,s) \wedge c(s,r,t) \wedge c(q,r,v) \wedge c(p,v,u) \Rightarrow t \approx u.$$

If a label token t is a concatenation of a label which is a concatenation of label tokens p and q and label token r, and a label token u is a concatenation of label p and a concatenation of labels q and r, then the labels t and u are equiform.

AXIOM I.6

$$c(p,q,t) \wedge c(r,s,t) \Rightarrow (p \approx r \Leftrightarrow q \approx s).$$

If one and the same label token is, on the one hand, a concatenation of labels p and q and, on the other, a concatenation of labels r and s, then p and r are equiform if and only if q and s are equiform.

AXIOM I.7

$$c(p,q,r) \wedge s \approx r \Rightarrow c(p,q,s).$$

A label which is equiform with a label token which is a concatenation of two labels is also a concatenation of those labels.

AXIOM I.8

$$c(p,q,t) \wedge c(r,s,u) \wedge t \approx u \Rightarrow p \approx r \vee \bigvee_{v} (c(r,v,p) \vee c(p,v,r)).$$

If two equiform label tokens are concatenations of pairs of label tokens with first elements p and r, respectively, then p and r are equiform or at least one of them is a concatenation of the other and of some label token.

The explanation of the intuitive sense of these axioms can, it seems, be left without a commentary. We shall give by way of example several theorems that follow from the axioms adopted above.

Axioms I.3 and I.1a yield immediately

CONCLUSION I.1

$$\neg \bigvee_{q} c(p,q,p) \wedge \neg \bigvee_{q} c(q,p,p).$$

No label token is a concatenation of itself and some label token nor a concatenation of some label token and itself.

This conclusion lets us note that the existence of. empty labels is not assumed in **TLTk**, unlike in the case of most theories of formal languages. The adoption of Axiom I.3, from which Conclusion I.1 follows, has no essential importance for the analyses carried out in this chapter, but is in agreement with intuition and is useful for reflections on the theory of syntactically categorial languages.

THEOREM I.1

$$c(p,q,u) \wedge p \approx r \wedge q \approx s \Rightarrow c(r,s,u).$$

A label token which is a concatenation of two labels is a concatenation of labels which are equiform with them.

To prove this theorem it suffices to note that by Axiom I.2 for the labels r and s there is a label r_1 such that $c(r,s,r_1)$, hence, under the assumption of this theorem and by Axiom I.4, $u \approx r_1$, and further, by Axiom I.7, $c(r,s,u)$, which was to be demonstrated.

Axioms I.6 and I.7 yield

CONCLUSION I.2

$$c(p,q,t) \wedge c(r,s,u) \wedge t \approx u \Rightarrow (p \approx r \Leftrightarrow q \approx s).$$

If equiform label tokens are concatenations of pairs of label tokens p and q and r and s, respectively, then p and r are equiform if and only if q and s are equiform.

We shall now record one more theorem, to be used further in the text.

THEOREM I.2

$$c(p,q,t) \wedge c(r,s,u) \Rightarrow (t \approx u \Leftrightarrow p \approx r \wedge q \approx s \vee \vee \bigvee_v (c(r,v,p) \wedge c(v,q,s) \vee c(p,v,r) \wedge c(v,s,q))).$$

Two label tokens which are concatenations of pairs of label tokens p and q and r and s, respectively, are equiform if and only if either the first elements of those pairs, i.e., p and r, are equiform and the second elements of those pairs, i.e., q and s, are equiform, or at least one of the first elements, i.e., p or r, is a concatenation of the other and a certain label, while the second element of the pair, i.e., s or q, respectively, is a concatenation of that label and the remaining label q or s.

The rather difficult proof of this theorem is given in the Annex. In that proof use is made of Axioms I.1, I.3, I.5, I.6, and I.8.

Sec. I.4. Vocabulary

The *vocabulary V* - the last primitive concept of **TLTk** - is the point of departure in the description of a given language. It is postulated about that set that it satisfies in particular the following axioms:

AXIOM I.9

$$\emptyset \neq V \subseteq Lb.$$

The vocabulary is a non-empty subset of the set of all label tokens.

AXIOM I.10

$$p \in V \wedge q \approx p \Rightarrow q \in V.$$

A label which is equiform with a label included in the vocabulary is also included in the vocabulary.

The vocabulary of formal languages is also often called *alphabet* or *set of letters* (see, e.g., A.A.Markov [1954],

M.Gross, A.Lentin [1967], J.E.Hopcroft, J.Ullman [1969]).

Elements of the vocabulary are called ***simple word tokens***. Simple words cannot have as parts (in the sense of Leśniewski's mereology) words that are different from them. Simple words are used to form, as their concatenations, compound words and expressions of a given language. On the contrary, simple words as such are not concatenations of any label tokens. This is provided for by the following axiom:

Axiom I.11

$$c(p,q,r) \Rightarrow r \notin V.$$

Concatenation of two labels is not a simple word.

The properties which simple word tokens have may not be properties of any labels. Note that Axiom I.11 is true for simple word tokens, but is not true for just any label tokens.

Theorem I.3 describes some properties of simple words and confirms the comment made above.

Theorem I.3

$$c(p,q,t) \wedge c(r,s,t) \wedge (p,r \in V \vee q,s \in V) \Rightarrow p \approx r \wedge q \approx s.$$

If a label token is a concatenation of two pairs of labels such that their first or second elements are simple word tokens, then the first and the second elements of those pairs are equiform.

Theorem I.3 follows easily from Theorem I.2 and Axioms I.1a and I.11.

Sec. I.5. Word tokens

A word (in Eng. also string; in Ger. Wort; in Fr. mot) of a given language need not of course be an element of the vocabulary. Those *word tokens* in which we are interested here are in the superset W of the vocabulary V. The set W - the *set of all word tokens* - is a secondary concept in **TLTk**. Its definition will be preceded by a convention and an auxiliary definition.

C o n v e n t i o n I.1b. The letter X, with or without subscript, is a variable which ranges over the family 2^{Lb}.

DEFINITION I.1a

$$\lambda(X) \Leftrightarrow V \subseteq X \wedge \bigwedge_{p,q,r}(p,q \in X \wedge c(p,q,r) \Rightarrow r \in X).$$

A set X satisfies the sentential expression $\lambda(X)$ *if and only if X contains the vocabulary V and is closed under the relation of concatenation.*

The definition of W is as follows:

DEFINITION I.1

$$W = \bigcap \{ X \mid \lambda(X) \}.$$

The set of all word tokens is the least set of label tokens which satisfies the expression $\lambda(X)$.

Note that - under Axiom I.9, Conventions I.1a,b, and Definition I.1a - Lb satisfies the condition $\lambda(X)$. Thus in accordance with both definitions W is the least subset of the set of all label tokens which contains V and is closed under the relation of concatenation c. This and the well-known theorems on the least sets which contain a given set and are closed under a given relation (see, e.g., L.Borkowski [1970], p.222-4) yield

CONCLUSION I.3a

$$V \subseteq W \subseteq Lb.$$

The set of all word tokens contains the vocabulary and is a subset of the set of all label tokens.

CONCLUSION I.3b

$$p,q \in W \wedge c(p,q,r) \Rightarrow r \in W.$$

A label which is a concatenation of two word tokens is also a word token.

CONCLUSION I.3c

$$V \subseteq W \wedge \bigwedge_{p,q,r} (p,q \in X \wedge c(p,q,r) \Rightarrow r \in X) \Rightarrow W \subseteq X.$$

W is contained in any set of label tokens which contains the vocabulary and is closed under the relation of concatenation.

W and Lb are non-empty sets under Axiom I.9 and Conclusion I.3a. Hence the set $W \setminus V$ is non-empty, too, because if p is an element of the non-empty set W, then under Axiom I.2 there is an r_1 such that $c(p,p,r_1)$ and under Conclusion I.3b and Axiom I.11 $r_1 \in W \setminus V$. This yields

CONCLUSION I.4

a. $W \neq \emptyset \wedge Lb \neq \emptyset,$

b. $W \setminus V \neq \emptyset.$

Elements of $W \setminus V$ will be termed *compound word tokens*.

Two more axioms hold in **TLTk**. They are axioms which describe the properties of W.

AXIOM I.12

$$r \in W \setminus V \Rightarrow \bigvee_{p,q \in W} c(p,q,r).$$

Every compound work token is a concatenation of a pair of word tokens.

AXIOM I.13

$$r \in W \wedge c(p,q,r) \Rightarrow p,q \in W.$$

If a word token is a concatenation of two label tokens, then those label tokens are word tokens.

We shall now give several theorems that follow from the axioms adopted above. Axioms I.12 and I.11 and Conclusion I.3b yield

THEOREM I.4a

$$r \in W \setminus V \Leftrightarrow \bigvee_{p,q \in W} c(p,q,r).$$

A compound word token is a label which is a concatenation of two word tokens.

THEOREM I.4

$$r \in W \Leftrightarrow r \in V \vee \bigvee_{p,q \in W} c(p,q,r).$$

A label token is a word token if and only if it is either a simple word or a concatenation of some two word tokens.

Theorem I.4 follows directly from Theorems I.4a and Conclusion I.3a.

Conclusion I.3a, Axiom I.11 and Theorem I.4a yield

THEOREM I.4b

$$r \in V \Leftrightarrow r \in W \wedge \neg \bigvee_{p,q \in W} c(p,q,r).$$

A label token is a simple word if and only if it is a word token which is not a concatenation of any pair of word tokens.

Theorem I.4a and Axioms I.10, I.7 and Conclusion I.3a yield

THEOREM I.5

$$p \in W \wedge q \approx p \Rightarrow q \in W.$$

A label token which is equiform with a word token is also a word token.

From Axiom I.13 and Conclusion I.3b we obtain immediately

THEOREM I.6

$$c(p,q,r) \Rightarrow (r \in W \Leftrightarrow p,q \in W).$$

A label token which is a concatenation of two label tokens is a word token if and only if those labels are word tokens.

In the proof of theorems of **TLTk** it is sometimes convenient to avail oneself of a certain metatheorem of that theory, which is, as it were, an analogon of the principle of induction. We adopt the convention

C o n v e n t i o n I.2. Theses (accepted sentences) whose numbers are marked with an asterisk are theses of the metatheory **TLTk**,

and record our metatheorem thus:

THEOREM I.1*. *Let $T(p)$ be a sentential expression of the language of* **TLTk** *recorded with the use of a free variable p and such that free variables q and r do not occur in it. Let $T(q)$, $T(r)$ be expressions obtained from $T(p)$ by a proper substitution in it of q and r for p. Then, if any simple word p satisfies $T(p)$ and if for any word tokens p and q which satisfy $T(p)$ and $T(q)$, respectively, their concatenation satisfies $T(r)$, then any word token p satisfies $T(p)$. In symbols*:

$$\bigwedge_{p\in V} T(p) \wedge \bigwedge_{p,q\in W} \bigwedge_{r} (T(p) \wedge T(q) \wedge c(p,q,r) \Rightarrow T(r)) \Rightarrow \bigwedge_{p\in W} T(p).$$

P r o o f. Assume that

$$(1) \qquad \bigwedge_{p\in V} T(p)$$

and that

$$(2) \qquad \bigwedge_{p,q\in W} \bigwedge_{r} (T(p) \wedge T(q) \wedge c(p,q,r) \Rightarrow T(r)).$$

Let W_1 be the set of all those words which satisfy $T(p)$, that is,

$$(3) \qquad W_1 = \{ p \in W \mid T(p) \}.$$

We shall demonstrate that W_1 is a set which contains the vocabulary V and is closed under the relation of concatenation. In fact, it follows from assumption (1), Conclusion I.3a and formula (3) that

$$(4) \qquad V \subseteq W_1.$$

On the other hand, if $p,q \in W_1 \wedge c(p,q,r)$, then the definition (3) of W_1 and formula (2) yield $T(r)$, where $r\in W$ in accordance with Conclusion I.3b. Hence, in view of formula (3), $r \in W_1$. This shows that W_1 is closed under the relation of concatenation, i.e.,

(5) $$\bigwedge_{p,q,r} (p,q \in W_1 \wedge c(p,q,r) \Rightarrow r \in W_1).$$

From formulas (4) and (5), under Conclusion I.3c, we obtain

$$W \subseteq W_1,$$

which, in accordance with (3), yields the thesis of the theorem being proved.

Theorem I.1* may be termed the principle of induction for the least set which contains the vocabulary and is closed under the relation of concatenation, and hence the *principle of induction for the set W*.

Sec. I.6. *n*-componential words; length of word

Compound words of a given language are generated from the set of simple words, which follows from Definition I.1 of W. This is so because any compound word is recorded exclusively with the use of those initial words. The "structure" of words is thus determined by their component parts (in the sense of Leśniewski's mereology), which are simple words. Before we define the concept of finitely componential word we refer to the usually adopted convention:

C o n v e n t i o n I.3. Letters i, j, k, l, m, n,..., with or without subscripts, are variables which range over the set $N\setminus\{0\}$ of natural numbers without zero.

We shall now introduce the definition of the concept of the

set ${}^{n}V$ *of all n-componential word tokens.* Elements of that set will be called *n-componential word tokens.* In formal languages they are usually termed "*n*-letter words". The set ${}^{n}V$ is defined by induction as follows (see G.Bryll, S.Miklos [1977]):

DEFINITION I.2

a. $\quad {}^{1}V = V,$

b. $\quad r \in {}^{n+1}V \Leftrightarrow \bigvee_{p\in {}^{n}V} \bigvee_{q\in V} c(p,q,r).$

A one-componential word token is a simple word; an n+1-componential word token is a label which is a concatenation of an n-componential word token and a simple word token.

We shall now formulate several theorems which result from the definition adopted above and describe the properties of *n*-componential word tokens. Their inductive proofs are given in the Annex.

Note first that the following theorem holds:

THEOREM I.7

$${}^{n}V \subseteq W.$$

Every n-componential word token is a word token.

Theorem I.5 has its analogon in

THEOREM I.8

$$p \in {}^{n}V \wedge q \approx p \Rightarrow q \in {}^{n}V.$$

A word token which is equiform with an n-componential word token is also an n-componential word token.

Two other theorems follow:

Theorem I.9

$$p \in {}^{m}V \wedge q \in {}^{n}V \wedge c(p,q,r) \Rightarrow r \in {}^{m+n}V.$$

If a label token is a concatenation of two words, m-componential one and n-componential one, then that label is an m+n-componential word.

Theorem I.10

$$r \in {}^{n+1}V \Leftrightarrow \bigvee_{p\in V} \bigvee_{q\in {}^{n}V} c(p,q,r).$$

A label token is an n+1-componential word token if and only if that label is a concatenation of a simple word token and an n-componential word token.

Theorem I.10 can replace Definition I.2b above, because, as can easily be shown, that definition follows from this theorem. It is not essential whether an $n+1$-componential word token is obtained as a concatenation of an n-componential word and a simple word or a simple word and an n-componential word.

We shall now formulate the theorem which refers to the comment made in the initial part of this Section.

Theorem I.11a

$$p \in W \Rightarrow \bigvee_{n} p \in {}^{n}V.$$

Every word token is a finitely componential word.

In the proof of this theorem we avail ourselves, by way of example, of Theorem I.1*, that is the principle of induction for W.

Note that the present theorem is true for any $p \in V$, which fact follows from Definition I.2a. Let us now assume "by induction" that it is true for word tokens ρ_1 and ρ_2 and show that it is true for the word r which is their concatenation. It follows from the assumption that $\rho_1 \in {}^{n_1}V$ and $\rho_2 \in {}^{n_2}V$, and hence by Theorem I.9 we have that $\bigvee_n r \in {}^nV$. By applying Theorem I.1* we find that the theorem being proved is true for any word token.

Theorems I.11a and I.7 yield

Theorem I.11

$$W = \bigcup_{k=1}^{\infty} {}^kV.$$

The set of all word tokens is the set of all finitely componential word tokens.

Theorems I.11 and I.11a have shown that every word token is in a set of finitely componential words. Is the membership in such a set determined unambiguously? The answer in the affirmative is provided by

Theorem I.11b

$$\rho \in W \Rightarrow \bigvee_n^1 \rho \in {}^nV.$$

Every word token is a word which is a member of precisely one set of finitely componential words.

The proof is given in the Annex.

By this theorem an n-componential word cannot have "componentiality" other than n, and hence cannot be, e.g., an m-componential word if $m \neq n$.

The rather clumsy word "componentiality" will be replaced by the word "length". In accordance with the symbolism most commonly used in the literature of the subject the *length of a word token* p will be recorded as $|p|$. In **TLTk** the definition of that concept takes on the following form:

DEFINITION I.3

$$p \in W \Rightarrow (|p| = n \Leftrightarrow p \in {}^{n}V).$$

A given word token has the length n if and only if it is an n-componential word.

The correctness of this definition is guaranteed by Theorem I.11b.

Note that Conclusion I.3a, Definitions I.3 and I.2a yield

CONCLUSION I.5a

$$p \in V \Leftrightarrow p \in W \wedge |p| = 1.$$

A simple word token is a word token of length 1.

Note further that Theorem I.11a and Definitions I.2a and I.3 imply

CONCLUSION I.5b

$$p \in W \setminus V \Leftrightarrow \bigvee_{n} (p \in W \wedge |p| = n > 1).$$

A compound word token is a word token of a length which is a natural number greater than 1.

Further, Definition I.3 and Theorems I.5, I.11a and I.8 yield

CONCLUSION I.6

$$p \in W \wedge q \approx p \Rightarrow |q| = |p|.$$

A word which is a equiform with a given word token has the same length as the latter.

Definition I.3 and Theorems I.11a, I.9 and I.7 yield

CONCLUSION I.7

$$p,q \in W \wedge c(p,q,r) \Rightarrow |r| = |p| + |q|.$$

The concatenation of two word tokens has the length equal to the sum of the lengths of those words.

The length of a word token is always a natural number other than zero, which follows from Definition I.3, Theorem I.11b and Convention I.3. This and Conclusion I.7 yield accordingly

CONCLUSION I.8

$$p,q \in W \wedge c(p,q,r) \Rightarrow |r| > |p| \wedge |r| > |q|.$$

The concatenation of two words has length which is greater than the length of either of those words.

Conclusion I.7 also yields immediately

CONCLUSION I.9

$$p,q \in W \wedge c(p,q,r) \wedge c(q,p,t) \Rightarrow |r| = |t|.$$

Concatenations of word tokens p and q and q and p have the same length.

Theorem I.4a and Conclusion I.7 yield

CONCLUSION I.10

$$r \in W \setminus V \Leftrightarrow \bigvee_{p,q \in W} (c(p,q,r) \wedge |r| = |p| + |q|).$$

A compound word token is a word which is a concatenation of two word tokens such that the sum of their lengths equals the length of that word.

We shall also give one more theorem which describes words of a definite length > 1.

THEOREM I.12

$$r \in W \wedge |r| = n > 1 \Leftrightarrow \bigwedge_{\substack{k,l \\ k+l=n}} \bigvee_{p,q \in W} (|p| = k \wedge |q| = l \wedge c(p,q,r)).$$

A word token of length $n > 1$ is a word which is a concatenation of each pair of word tokens such that the sum of their lengths equals n.

Sec. I.7. Generalized concatenation

In Sec.I.6 we have formulated a theorem (Theorem I.11a) which states that word tokens are finitely componential words. But the definitions and theorems formulated so far do not allow us to state the intuitively obvious fact that an n-componential word, for $n > 1$, that is a word of a length $n > 1$, is a label obtained from the concatenation of n simple word tokens. In order formally to write down the theorem just formulated above we shall make use of the term c^{n+1}, which denotes the *generalized relation of concatenation* of $n+2$ arguments.

The expression $c^{n+1}(p_1,p_2,\ldots,p_{n+1},p)$ is read: p is a concatenation of $n+1$ elements, namely of successive label tokens $p_1, p_2, \ldots, p_{n+1}$. The labels $p_1, p_2, \ldots, p_{n+1}$ are then said to be the successive *components* of that concatenation.

Generalized relation of concatenation c^{n+1} $(n \geq 1)$ is defined by induction.

DEFINITION I.4

a. $n = 1 \Rightarrow (c^{n+1}(p_1, p_2, p) \Leftrightarrow c(p_1, p_2, p)),$

b. $n > 1 \Rightarrow (c^{n+1}(p_1, p_2, \ldots, p_{n+1}, p) \Leftrightarrow$

$$\Leftrightarrow \bigvee_q (c^n(p_1, p_2, \ldots, p_n, q) \wedge c(q, p_{n+1}, p))).$$

In accordance with this definition p is a two-element concatenation of two label tokens means the same as p is an "ordinary" concatenation of those two labels; p is an $n+1$-element $(n>1)$ concatenation of a label tokens $p_1, p_2, \ldots, p_n, p_{n+1}$ if and only if it is a concatenation of label which is an n-element $(n>1)$ concatenation of labels $p_1, p_2, \ldots, p_n$ and a label p_{n+1}.

And here is the announced

THEOREM I.13a

$$p \in W \wedge |p| = n+1 \Leftrightarrow \bigvee_{p_1, p_2, \ldots, p_{n+1} \in V} c^{n+1}(p_1, p_2, \ldots, p_{n+1}, p).$$

A word token of length n+1 is a label which is an n+1-element concatenation of certain simple word tokens.

In accordance with Conclusion I.5b, Theorem I.13a makes it possible to formulate the following conclusion: a compound word token is a word which is a concatenation of a finite number of componential simple word tokens, and the number of those componential simple words equals the length of the word in question.

The proof of Theorem I.13a is given in the Annex.

The following theorem also holds:

THEOREM I.13b

$$p \in W\setminus V \Leftrightarrow \bigvee_{n} \bigvee_{p_1, p_2, \ldots, p_{n+1} \in W} c^{n+1}(p_1, p_2, \ldots, p_{n+1}, p).$$

A compound word token is a label which is a certain concatenation of a finite number of word tokens.

The simple implication in Theorem I.13b follows from Conclusion I.5b, Theorem I.13a, and Conclusion I.3a. To prove the converse implication it suffices to refer to

LEMMA I.1

$$\bigwedge_{1 \le i \le n+1} p_i \in W \wedge c^{n+1}(p_1, p_2, \ldots, p_{n+1}, p) \Rightarrow p \in W\setminus V.$$

This lemma is a generalization of Conclusion I.3b. Its proof by induction, based on that conclusion, Axiom I.11 and Definition I.4a,b, is left to the Reader.

Conclusions I.3a, I.5b and Theorem I.13a yield one more theorem (next to Theorems I.4 and I.11) which describes the set W of all words:

Theorem I.13

$$p \in W \Leftrightarrow p \in V \vee \bigvee_{p_1, p_2, \ldots, p_{n+1} \in V} c^{n+1}(p_1, p_2, \ldots, p_{n+1}, p).$$

A word token is either a simple word token or a label which is a concatenation of a finite number of simple word tokens.

Simple word tokens which are components of a given compound word are, up to equiformity, determined unambiguously. This is so because there holds a generalization of the theorem

$$c(p,q,t) \wedge c(r,s,t) \wedge p,r,q,s \in V \Rightarrow p \approx r \wedge q \approx s,$$

which is a weakened form of Theorem I.3:

Theorem I.14

$$c^{n+1}(p_1, p_2, \ldots, p_{n+1}, p) \wedge c^{n+1}(q_1, q_2, \ldots, q_{n+1}, p) \wedge$$

$$\wedge \bigwedge_{1 \le i \le n+1} p_i, q_i \in V \Rightarrow \bigwedge_{1 \le i \le n+1} p_i \approx q_i.$$

If one and the same word token can be presented as a generalized concatenation of two configurations of simple word tokens, then the components of those configurations occurring in the same places are equiform.

The proof of Theorem I.14 is given in the Annex.

The concept of generalized concatenation has a considerable practical importance, being useful in analyses to be found later in this work. For our purposes its practical usefulness is to be seen in particular in certain theorems which are generalizations of theorems formulated for the "ordinary" concatenation.

The following theorem is a generalization of Axiom I.7; it

follows from that axiom and Definition I.4a,b.

THEOREM I.15

$$c^{n+1}(p_1,p_2,\dots,p_{n+1},p) \wedge q \approx p \Rightarrow c^{n+1}(p_1,p_2,\dots,p_{n+1},q).$$

A label token which is equiform with a label token that is a concatenation of n+1 label tokens is also a concatenation of n+1 those labels.

Theorem I.16 is a generalization of Theorem I.1; it follows from the latter and Definition I.4a,b.

THEOREM I.16

$$c^{n+1}(p_1,p_2,\dots,p_{n+1},p) \wedge \bigwedge_{1\le i\le n+1} q_i \approx p_i \Rightarrow$$

$$\Rightarrow c^{n+1}(q_1,q_2,\dots,q_{n+1},p).$$

If a given label token is a concatenation of n+1 label tokens, then it is also a concatenation of n+1 successive labels which are pairwise equiform with the successive components of the initial concatenation.

The simple proofs of Theorems I.15 and I.16 are left to the Reader.

The two above theorems yield immediately

CONCLUSION I.11

$$c^{n+1}(p_1,p_2,\dots,p_{n+1},p_{n+2}) \wedge \bigwedge_{1\le i\le n+2} p_i \approx q_i \Rightarrow$$

$$\Rightarrow c^{n+1}(q_1,q_2,\dots,q_{n+1},q_{n+2}),$$

in accordance with which if the relation of concatenation of

$n+2$ arguments holds between given label tokens, then it also holds between successive labels which are pairwise equiform with the elements of the first concatenation.

Axiom I.4 has its inductive generalization in

THEOREM I.17

$$c^{n+1}(p_1,p_2,\ldots,p_{n+1},p) \wedge c^{n+1}(q_1,q_2,\ldots,q_{n+1},q) \wedge$$
$$\wedge \bigwedge_{1\leq i\leq n+1} p_i \approx q_i \Rightarrow p \approx q.$$

Two label tokens which are concatenations of n+1 labels each and are such that any of their i-th components (1≤i≤n+1) are pairwise equiform, are equiform.

The easy proof of this theorem is omitted.

To conclude this section we formulate a theorem which is a generalization of Conclusion I.2:

THEOREM I.18

$$c^{n+1}(p_1,p_2,\ldots,p_{n+1},p) \wedge c^{n+1}(q_1,q_2,\ldots,q_{n+1},q) \wedge$$
$$\wedge\ p \approx q \wedge 1 \leq j \leq n+1 \wedge \bigwedge_{\substack{1\leq i\leq n+1 \\ i\neq j}} (p_i \approx q_i) \Rightarrow p_j \approx q_j.$$

If two equiform label tokens are n+1-element concatenations of two configurations such that all components occurring in the same place, except for the j-th components (1≤j≤n+1), are pairwise equiform, then those j-th components are equiform, too.

The proof of Theorem I.18 is given in the Annex.

Sec. I.8. Methodological remarks

The theory **TLTk**, presented in the preceding sections and based on the system of primitive concepts

(α) $Lb, \approx, c, V$

is equivalent to theory of label tokens outlined in the papers [1981a,1983a] by the present author and modelled on the theory of concrete and abstract words formulated by G.Bryll and S.Miklos [1977]; like the latter theory, it was based on the system of primitive concepts

(β) $Lb, \approx, c, W$

which differed from (α) by having the set of all word tokens W instead of the vocabulary V; the latter is defined by the formula

(1) $$r \in V \Leftrightarrow r \in W \wedge \neg \bigvee_{p,q \in W} c(p,q,r).$$

The definitions of the following concepts: the set ${}^{n}V$, length $|p|$, and the relation c^{n+1} are identical in both theories. Axioms I.1–I.8 are identical, too, and Axioms I.9–-I.13 of **TLTk** are replaced by the axioms

(2) $$\emptyset \neq V \subseteq W \subseteq Lb,$$

(3) $$p \in W \wedge q \approx p \Rightarrow q \in W,$$

(4) $$V \subseteq X \wedge \bigwedge_{p,q \in X} \bigwedge_{r} (c(p,q,r) \Rightarrow r \in X) \Rightarrow W \subseteq X,$$

(5) $$c(p,q,r) \Rightarrow (r \in W \Leftrightarrow p,q \in W).$$

Thus the following theorem holds:

THEOREM I.2*. *The theory of label tokens based on the system* (β) *of primitive concepts, on the axiom system consisting of Axioms I.1–I.8 and expressions* (2)–(5), *and on Definition* (1) *is equivalent to the theory* **TLTk**.

P r o o f. Note first that the theory based on Definition (1) and the axiom systems consisting of Axioms I.1–I.8 and expressions (2)–(5) is contained in **TLTk**, because the expression (1) is Theorem I.4b in **TLTk**, the expression (2) follows immediately from Axiom I.9 and Conclusion I.3a, and the expressions (3)–(5) are Theorem I.5, Conclusion I.3c, Theorem I.6, respectively. The converse inclusion also holds. Definition I.1 of the set W adopted in **TLTk** is a theorem in the theory based on Definition (1) and the axiom system consisting of Axioms I.1–I.8 and (2)–(5), because if $r \in W$, then - if we adopt the definition of the expression $\lambda(X)$ the same as in **TLTk** (Definition I.1a) - on the strength of Condition (4) for every set Y which satisfies $\lambda(X)$, $r \in Y$, and hence $W \subseteq \bigcap \{X \mid \lambda(X)\}$. On the other hand, if $r \in \bigcap \{X \mid \lambda(X)\}$, which is to say that $r \in Y$ for any set Y which satisfies $\lambda(X)$, then $r \in W$, because W satisfies $\lambda(X)$, which follows from Conditions (2) and (5). Axioms I.9–I.13 of **TLTk** are also theorems in the theory based on the (β) system of primitive concepts, because Axiom I.9 follows directly from Condition (2), Axioms I.11 and I.12 follow directly from Definition (1), Axiom I.13 follows immediately from Condition (5), and finally Axiom I.10 follows easily from

Conditions (2) and (3), and Axioms I.12, I.7, and I.11.

In the paper [1983a] by the present author it was demonstrated that the theory of label tokens based on Definition (1) and Axioms I.1–I.8, (2)–(5) is consistent. Theorem I.2* thus yields

THEOREM I.3*. *The theory* **TLTk** *is consistent.*

The consistence of **TLTk** is also immediately obtained by its interpretation in the arithmetic of natural numbers if Lb and W are interpreted as the set N of natural numbers without zero, the set V, as the set {1}, the only element of which is the number 1, the relation $\approx$, as identity, and the relation c as the function of addition.

CHAPTER II

THE AXIOMATIC SYSTEM TSCL OF SIMPLE CATEGORIAL LANGUAGES

Sec. II.1. The principal objectives of the construction of the theory of syntactically categorial languages

The discussions of the theory **TSCL** [3] of simple categorial languages, which is to be constructed in this chapter, will be concerned with an arbitrary but fixed language $\mathcal{L}$ of expression tokens which do not include variables bound by operators. The language $\mathcal{L}$, called here a *simple categorial language* of expression tokens, is determined by certain conditions imposed on its syntax in accordance with the idea of K.Ajdukiewicz's categorial grammar [1935] and the conceptions of those researchers who referred to his ideas.[4]

In Sec.I.1 we gave a simplified characteristic of any language in the form of the sixtuple

$$(\mathcal{L}) \qquad \langle\, Lb;\ V,\ W;\ \approx,\ \mathrm{c};\ S\,\rangle$$

which consists, successively, of the set of all label tokens, the vocabulary, the set of all word tokens, the relation of equiformity, the relation of concatenation, and the set of all meaningful, i.e., well-formed expression tokens. All the elements of $(\mathcal{L})$, except the last one, are described to be the theory **TLTk**, presented in Chap.I. The set S of all well-formed expression tokens is a concept which is defined in the theory

TSCL and also (see Chap.III) in the theory **TSCω-L**, pertaining to the ***categorial ω-languages***, i.e., languages of expression tokens which may include ***operators*** and ***variables*** bound by them. The definitions of S require the introduction into both theories of the set of categorial indices (see Secs.II.2 and II.3), because categorial indices are the foundation of all categorial languages. Although they perform an auxiliary function, their role is essential not only because they are used in the definition of the concept of a well-formed expression (see Secs.II.5 and III.4) and are useful in the checking of syntactic correctness, that is the syntactic connexion of expressions of a given language (see Secs.II.8 and III.6). They are used above all in indicating the syntactic categories of the expressions of a given language and enable us to carry out a logical classification $Ct(S)$ of S into syntactic categories (see Secs.II.6 and III.5).

The set I of all categorial indices is generated from the set I_o *of all basic indices*. The basic indices which are elements of I_o are used to indicate the basic categories of expressions. The remaining indices determine the derivative, functoral, categories.

Indices are univocally assigned to definite word tokens of a given syntactically categorial language by means of a certain function termed the *function of indication of indices* and symbolized by ι. The set E of all expression token of a given language and its subset S will be selected from the set of words that have indices, that is, from the domain of the function ι.

A certain syntactic operation ρ, which makes it possible to put together, and hence combine by the appropriate concatenation, any finite, ordered configuration of word tokens that have indices into a compound word token that has an index and is a compound expression token of the language, will be useful in defining the set E. For instance, in the contemporary descriptions of the language of sentential calculus, whose vocabulary includes round brackets, the operation ρ has its analogon in an operation which is particular to any two word tokens ϕ_0 and ϕ_1 of the language of that calculus, which have an index, assigns the four-element concatenation $\phi_0(\phi_1)$, and to any three word tokens ϕ_0, ϕ_1 and ϕ_2, which have an index, assigns the seven-element concatenation $(\phi_1)\phi_0(\phi_2)$. It is obvious that not every expression formed in this way is a well-formed expression of the language of sentential calculus. If well-formedness is to be assured, definite conditions must be imposed upon the indices of the words.

The remarks made above show that the syntactic characteristic of any syntactically categorial language in the form of $(\mathscr{L})$ is greatly simplified and requires completion. The simple categorial language $\mathscr{L}$, which is the foundation of the theory TSCL constructed in the present chapter, is characterized by a much more complex system, namely

$$(\mathscr{L}_c) \qquad \langle\, Lb;\ V,\ W;\ \approx,\ c;\ I_o,\ I;\ \iota,\ \rho;\ E,\ S;\ Ct(S)\,\rangle,$$

whose elements satisfy certain assumptions postulated in that theory.

It is worth noting that the syntactic characterization of any ω-language, to which the theory **TSCω-L** pertains, has the form of an ordered system much more complex than ($\mathscr{L}_c$) (see Sec.III.1).

The analyses carried out in **TSCL** and **TSCω-L** clearly refer to the theory of syntactic categories and result (see Secs.II.7, III.5) in two theorems which might by termed *fundamental theorems of the theory of syntactic categories*, abbreviated as **fttsc**.

Their formulation is based on the expression

(α) $$r(p/q)s,$$

whose corresponding definitions in **TSCL** and **TSCω-L** are important definitions in those theories.

The expression (α) is read: the expression r is obtained from the expression s by the replacement of its constituent q by the expression p.

Now (α) is the common assumption of both **fttsc**. One of those theorems has the following expression as its second assumption:

(β) r, s are expressions of the same syntactic category.

That theorem has its thesis in the form of the expression:

(γ) p, r are expressions of the same syntactic category.

Thus the schema of the theorem under consideration has the implication

I. $$\alpha \wedge \beta \Rightarrow \gamma.$$

The second **fttsc** has the implication

II. $$\alpha \wedge \gamma \Rightarrow \beta$$

as its schema.

Under the assumption (α) the two **fttsc** are converse theorems.

Now (β) has its concretization in the expression:

(β') r,s are expressions of the syntactic category of sentences.

The implications

(I') $$\alpha \wedge \beta' \Rightarrow \gamma$$

or

(III) $$\alpha \Rightarrow (\gamma \Leftrightarrow \beta')$$

are schemata of the definitions of (γ), often adopted in the theory of syntactic categories. The treatment of syntactic categories as sets of expressions replaceable in sentences requires the introduction into the theory of syntactic categories of both (α) and the concept of expression that belongs to the syntactic category of sentences, that is, the concept of sentence.

Including the concept of sentence in the primitive concepts of the theory of syntactic categories would preclude the possibility of defining it. It seems that it is the defining of the concept of sentence that should be one of the fundamental tasks of the theory of syntactic categories as applied to natural and symbolic languages. It seems also that the definition of sentence should satisfy the condition which enables us to formulate the algorithm of checking the syntactic connectedness of expressions, analogous to that given by Ajdukiewicz. But a definition of sentence which satisfies this condition, when (I') and (III) are treated as a schema of the definition of belonging of two expressions to the same syntactic category, would involve the danger of a vicious circle. In constructing the theory of syntactically categorial languages we try to overcome those difficulties. Our endeavour is treated as the essential goal of the theory **TSCL** and its certain modification, the theory **TSCω-L**, outlined in the next chapter. In Secs.IV.5 and IV.6 we discuss two more theories of syntactically categorial languages: **DTSCL** and **DTSCω-L**. They are constructed in accordance with the ideas which underlie **TSCL** and **TSCω-L** for the purpose of describing languages of expression types.

Sec. II.2. Connections between TLTk and TSCL

Now **TSCL** like **TLTK** is based on classical logic and set theory. Hence all logical rules, in particular the rule of substitution, are valid in **TSCL** because **TSCL** is superstructured over **TLTk**. Hence again

REMARK II.1. The constant symbols of the vocabulary of **TLTk**

$$Lb,\ V,\ W,\ \approx,\ \mathrm{c}$$

are constant symbols of the vocabulary of **TSCL**.

The variable symbols of the vocabulary of **TSCL** include

(v_1) p, q, r, s, t, u, v, with or without subscripts,

and

(v_2) X, with or without subscript;

(v_1) and (v_2) range, respectively (see Convention I.1a,b), over

Lb and the family 2^{Lb} of all its subsets.

The variables listed under (v_1) and (v_2) do not exhaust the list of the variable symbols of the vocabulary of the language of **TSCL** (see Sec.II.5, Convention II.3b).

We observe the principle (see Convention I.2) that

C o n v e n t i o n II.1. Those theses (accepted sentences) whose numbers are marked with an asterisk are theses in the metatheory of **TSCL**, and note correspondingly that

THEOREM II.1*. *Every thesis (accepted sentence) of* **TLTk** *is a thesis of* **TSCL**.

Both theories now under consideration pertain to word tokens. In both theories it is *Lb* which is the universal set being taken into account. From that set we single out not only the set *W*, but also the *set I of all categorial indices*; as we shall see (Sec.II.3, Theorem II.1), *I* has no common elements with *W*.

Now *I* is a defined concept in **TSCL**. Its definition is recorded with the use of a new primitive term I_o, specific of **TSCL**, which denotes the *set of all basic indices* (I_o includes at least one technical symbol).

The set *I* will be defined with reference to the set I_o in the following manner (cf. Definition I.1a):

DEFINITION II.1a

$$\lambda(X)^{d} \Leftrightarrow I_{o} \subseteq X \wedge \bigwedge_{p,q,r} (p,q \in X \wedge c(p,q,r) \Rightarrow r \in X).$$

The set of label tokens satisfies the expression $\lambda(X)^{d}$ *if and only if it includes the set* I_{o} *and is closed under the relation of concatenation.*

Hence we have (cf. Definition I.1)

DEFINITION II.1

$$I = \bigcap \{ X \mid \lambda(X)^{d} \}.$$

The set of all categorial indices is the least set of label tokens which satisfies the expression $\lambda(X)^{d}$.

By these definitions I is the least set of label tokens which includes I_{o} and is closed under the relation of concatenation.

The properties of I_{o} are described above all by the first three of the following five axioms of TSCL (cf. Axioms I.9–I.13):

AXIOM II.1

$$\emptyset \neq I_{o} \subseteq Lb.$$

The set of indices is non-empty set of label tokens.

AXIOM II.2

$$p \in I_{o} \wedge q \approx p \Rightarrow q \in I_{o}.$$

A label which is equiform with a basic index is also a basic index.

AXIOM II.3

$$c(p,q,r) \Rightarrow r \notin I_o.$$

A basic index is not a concatenation of any two label tokens.

AXIOM II.4

$$r \in I \setminus I_o \Rightarrow \bigvee_{p,q \in I} c(p,q,r).$$

A categorial index which is not basic is a concatenation of a pair of categorial indices.

AXIOM II.5

$$r \in I \wedge c(p,q,r) \Rightarrow p,q \in I.$$

If a categorial index is a concatenation of two labels, then those labels are also categorial indices.

We shall now introduce a definition by induction of the *set* ${}^nI$ *of all n-componential categorial indices* (cf. Definition I.2a,b).

DEFINITION II.2

a. $\qquad {}^1I = I_o,$

b. $\qquad r \in {}^{n+1}I \Leftrightarrow \bigvee_{p \in {}^nI} \bigvee_{q \in I_o} c(p,q,r).$

A one-componential categorial index is a basic index; an n+1-componential categorial index is a label token which is a concatenation of an n-componential categorial index and a basic index.

The syntactic analogies of Definitions II.1a, II.1, and II.2a,b above with, Definitions I.1a, I.1, and I.2a,b of **TLTk**, respectively, which are, successively, definitions of the expression $\lambda(X)$, the set W and the set ${}^{n}V$, are self-evident. The same applies to the analogies between Axioms II.1–II.5 and the corresponding Axioms I.9–I.13 of **TLTk**. We shall formulate those observations more precisely by making use of the concept of a *dual analogon of an expression* of **TLTk**.

DEFINITION II.1*. *If ϕ is any expression of the language of* **TLTk** *which next to variables may include the symbol Lb and the symbols*

(1) $V,\ W,\ {}^{n}W,$

then the expression ϕ^{d} is a dual analogon of ϕ if and only if ϕ^{d} is obtained from ϕ by the replacement, in the latter, in each place of symbols (1) respectively by symbols

$(1)^{d}$ $I_{o},\ I,\ {}^{n}I.$

It is self-evident that

REMARK II.2. Definitions II.1a, II.1, and II.2a,b are dual analoga of Definitions I.1a, I.1, and I.2a,b, respectively. Axioms II.1–II.5 are dual analoga of Axioms I.9–I.13, respectively.

Note that, by Definition II.1*,

REMARK II.3. Axioms I.1a–c, I.2–I.8 have themselves as their dual analoga.

Theorem II.1* and Remarks II.3 and II.2 yield

THEOREM II.2*. *The dual analoga of all axioms and definitions accepted in* **TLTk** *are theses in* **TSCL**. [5]

Hence we have

THEOREM II.3*. *Every dual analogon of a thesis in* **TLTk** *is a thesis in* **TSCL**.

The practical values of Theorem II.3* can be shown if we list, by way of example, those theorems of **TSCL** which we shall use in the proof of the theorem given in Sec.II.3 below.

We adopt the following convention:

C o n v e n t i o n II.2. A dual analogon of a theorem of TLTk has the same number, but that number is marked by the superscript d.

Here are the theorems mentioned above:

CONCLUSION I.3^{d}a

$$I_o \subseteq I \subseteq Lb.$$

CONCLUSION I.4^{d}b

$$I \setminus I_o \neq \emptyset.$$

THEOREM I.10^d

$$r \in {}^{n+1}I \Leftrightarrow \bigvee_{p \in I_o} \bigvee_{q \in {}^nI} c(p,q,r).$$

THEOREM I.11^d

$$I = \bigcup_{k=1}^{\infty} {}^kI.$$

Sec. II.3. Categorial indices and their indication

Categorial indices were introduced into semiotics by Kazimierz Ajdukiewicz [1934,1935]. He used them in describing the hierarchy of semantic (syntactic) categories and in his investigations of the syntactic connectedness of linguistic expressions.

Categorial indices, by some students of categorial languages called just categories, may be objects of various kinds. It is often assumed that they are natural numbers.[6]

For the purposes of **TSCL** it is convenient to assume, as has been done earlier, that categorial indices are label tokens (cf. Definition II.1, Sec.II.2).

Axioms II.1–II.5, which are dual analoga of Axioms I.9–I.13 in **TLTk**, do not exhaust, of course, the list of those axioms which describe the properties of categorial indices. The next axiom is

AXIOM II.6

$$V \cap I_o = \emptyset.$$

The vocabulary V of a given categorial language $\mathcal{L}$ is disjoint with the set of all its basic indices.

Now I_o may be termed an *auxiliary vocabulary* relative to the vocabulary V of a given object language $\mathcal{L}$. From the set W (generated from the vocabulary V) of all word tokens (Definition I.1) we shall single out the set S of all well-formed expressions. In a similar way we can single out, from the set I of all indices (Definition II.1) (generated from the auxiliary vocabulary I_o), the set of *expressions indexing* the well-formed expressions of $\mathcal{L}$. Indexing expressions are, of course, expressions of a language other than $\mathcal{L}$, namely of a language which is part of a metalanguage relative to $\mathcal{L}$ (see

Sec.II.4).

The theorem announced in Sec.II.2 sheds much light upon the above remark:

THEOREM II.1

$$W \cap I = \emptyset.$$

The set of all word tokens is disjoint with the set of all categorial indices.

The proof is given in the Annex.

Categorial indices are assigned to the words of $\mathcal{L}$ by the *relation* ι *of indicating the indices of word tokens*. That relation is the second specific primitive concept in **TSCL**. It satisfies the following axioms:

AXIOM II.7

$$\iota \subseteq W \times I \textit{ and } \iota \textit{ is a function.}$$

ι is a function of one argument which maps a subset of W into a subset of I.

AXIOM II.8

$$p \in D(\iota) \wedge q \approx p \Rightarrow q \in D(\iota) \wedge \iota(q) \approx \iota(p).$$

A label token which is equiform with a word token that has an index is also a word that has an index which is equiform with the index of that word token.

The interpretation of Axiom II.8 becomes comprehensible if we note that the elements of the domain $D(\iota)$ of the function ι[7] are words to which an index has been assigned, while the elements of the counter-domain $D^{-1}(\iota)$ of that function are the indices of those words. This follows immediately from Axiom II.7 in the view of

CONCLUSION II.1

a. $D(\iota) \subseteq W$,

b. $D^{-1}(\iota) \subseteq I$.

The domain of ι is a subset of the set W of all word tokens while its counter-domain is a subset of the set I of all categorial indices.

Note that Axiom II.7 and Conclusion II.1 do not decide arbitrarily whether categorial indices are[8] or not are[9] assigned to structural, technical, signs usually included in the vocabulary of a given language (such as brackets); nor do they decide whether categorial indices are assigned only to the words of $\mathscr{L}$.

Conclusion II.1a seems natural also because in practice we are not inclined to assign a categorial index to every concatenation of the simple words of a given language. We do that, however, with reference to those words which are termed expressions (Ger. Ausdrücke) of that language. Expressions will be discussed in Sec.II.4.

Axioms II.8 and I.1b immediately yield

CONCLUSION II.2

a. $p \in D(\iota) \wedge q \approx p \Rightarrow q \in D(\iota)$,

b. $p,q \in D(\iota) \wedge q \approx p \Rightarrow \iota(p) \approx \iota(q)$.

We shall avail ourselves of Conclusion II.2b on many occasions.

Sec. II.4. Expressions

The one-to-one function ρ of one variable, called the *operation of forming compound expression tokens* is the third, and the last, primitive concept of **TSCL**. Its domain is the sum of all finite and greater than one Cartesian powers of the set $D(\iota)$ of all word tokens, which have a categorial index assigned to them. Its counter-domain is a subset of $D(\iota)\setminus V$. Hence we add the following axiom to **TSCL**:

AXIOM II.9

$$\rho: \bigcup_{k=2}^{\infty} D(\iota)^k \xrightarrow[1-1]{} D(\iota)\setminus V.$$

The compound word token denoted by

(δ) $\rho(p_0, p_1, \ldots, p_n)$,

which is the value of the operation ρ for the argument $\langle p_0, p_1, \ldots, p_n\rangle$ for $n \geq 1$ and is thus an element of the counter--domain of ρ, and hence also of the set $D(\iota)\setminus V$, is called a *compound expression token of* $\mathcal{L}$, *consisting of* $n+1$ ($n \geq 1$) *word tokens* $p_0, p_1, \ldots, p_n$, *in that order*. If there is no risk of misunderstanding, such an expression will be termed briefly a *compound expression of* $\mathcal{L}$.

To activate the appropriate intuitions connected with the compound expression (δ) we shall make certain remarks. They will not be precise, which is justified by their character. Their generality makes precision impossible.

We begin with several terminological conventions.

The sequence of word tokens

(ε) $p_0, p_1, \dots, p_n$

of which the compound expression (δ) of $\mathscr{L}$ is constructed will be called the *expansion of* (δ).

The word p_0 of the expansion (ε) will be called the *main functor of* (δ), and the words $p_1, p_2, \dots, p_n$ will be called the *arguments of* p_0, the word p_i ($1 \le i \le n$) - the *i-th argument* of that functor.

By analogy to the commonly used terms "sentence-forming functor" and "name-forming functor" we shall call the main functor of (δ) the *α-forming functor*, where *α* is the categorial index of (δ). The existence of that index follows from Axioms II.9 and II.7.

The label from the language of TSCL denoted by (δ) may be treated as a *translation* into the language of TSCL of such a compound expression of $\mathscr{L}$ which is constructed of the main functor p_0 and its successive arguments $p_1, p_2, \dots, p_n$. The term "translation" is left undefined; we merely assume that its intuitive sense agrees with that given to it by Alfred Tarski in his well-known monograph [1933].

In the present book the essential fact is that the translation of a compound expression of $\mathscr{L}$ into the language of TSCL does not depend upon the symbolism in which that expression is recorded; in particular, it is independent of whether a bracket notation is used in $\mathscr{L}$ or not. That translation depends merely on which words of the expansion of that compound expression or the expansion of its part (in the sense of Leśniewski's mereology) which is a compound expression are main functors and which are their arguments.

For instance, the same metalinguistic label is a translation of each of the following synonymous compound expressions of the language of the sentential calculus:

$$p \Rightarrow (q \Leftrightarrow r); \quad \phi\!-\!(p,\phi(q,r)); \quad \mathrm{CpEqr}.$$

The letters p,q,r are here, of course, sentential variables, and the expressions given above are written successively in the notation which is now most commonly used, Leśniewski's notation introduced in [1929], and Łukasiewicz's parenthesis-free notation, called Polish, [1929].

Axiom II.8 (see Conclusion II.2b) makes it possible to formulate an important property of compound expressions. The same applies to the following

AXIOM II.10

$$p = \rho(p_0,p_1,\ldots,p_n) \Rightarrow (q \approx p \Leftrightarrow$$

$$\Leftrightarrow \bigvee_{q_0,q_1,\ldots,q_n} (q = \rho(q_0,q_1,\ldots,q_n) \wedge \bigwedge_{0\le k\le n} q_k \approx p_k)).$$

If p is a compound expression consisting of n+1 (n≥1) successive word tokens, then a label token q is equiform with it if and only if it is also a compound expression consisting of n+1 word tokens and every word of the expansion of q is equiform with the word of the expansion of p occurring in the same place.

Axiom II.10 leads to

CONCLUSION II.3

$$p = \rho(p_0, p_1, \ldots, p_n) \wedge q = (q_0, q_1, \ldots, q_n) \Rightarrow$$
$$\Rightarrow (p \approx q \Leftrightarrow \bigwedge_{0 \le k \le n} p_k \approx q_k).$$

Two compound expressions consisting of the same number of word tokens each are equiform if and only if any two terms of their expansions which occur in the same place are equiform.

The proof is given in the Annex.

We formulate three further definitions:

DEFINITION II.3

a. $E_s = V \cap D(\iota)$,

b. $E_c = D^{-1}(\rho)$,

c. $E = E_s \cup E_c$.

The elements of the set E_s are called *simple expression tokens* of $\mathcal{L}$. The elements of the set E_c, i.e., of the counter-domain of the function ρ, have already earlier been termed *compound expression tokens* of $\mathcal{L}$.[10] The elements of the set E are called *expression tokens* of $\mathcal{L}$. Frequently, when speaking about the elements of E_s, E_c and E we shall just refer to simple expressions, compound expressions, and expressions. In accordance with Definition II.3c the set E of all expressions of $\mathcal{L}$ is the sum of the set of all its simple expressions and the set of all its compound expressions.

Definition II.3a, Axiom I.10, and Conclusion II.2a yield

CONCLUSION II.4a

$$p \in E_S \wedge q \approx p \Rightarrow q \in E_S.$$

A label token which is equiform with a simple expression is also a simple expression.

Definition II.3b, Axioms II.9 and II.10 and Conclusion II.2a yield

CONCLUSION II.4b

$$p \in E_C \wedge q \approx p \Rightarrow q \in E_C.$$

A label token which is equiform with a compound expression is also a compound expression.

Definition II.3c and Conclusions II.4a,b yield

CONCLUSION II.4c

$$p \in E \wedge q \approx p \Rightarrow q \in E.$$

A label token which is equiform with an expression of $\mathscr{L}$ is also an expression of $\mathscr{L}$.

Definitions II.3a,b,c and Axiom II.9 yield

CONCLUSION II.5

a. $E_C \subseteq D(\iota)\backslash V$,

b. $E \subseteq D(\iota)$.

The set of all compound expressions is included in the set of all compound words which have an index. The set of all expressions is included in the set of all words which have an index.

In accordance with Conclusion II.5b every expression has a categorial index assigned to it.

To conclude the discussion carried out in this section we want to point to a fragment of the theory TSCL now under construction, the fragment which describes the system of the form:

(E_C) $\langle Lb; V, W; \approx, c; I_O, I, \iota, \rho; E \rangle$.

In doing so we refer to the axioms and definitions of **TLTk** and Axioms II.1–II.10 and Definitions II.1a and II.1–II.3. That fragment also includes Definition II.8–II.10, II.12 and II.13.[11] The fragment now under consideration will be called the *theory of expression tokens* and denoted by **TETk**.

TETk is the foundation of **TSCL** and also of the theory **TSCω-L**, analysed in the next chapter and pertaining to *ω-languages*, that is languages with operators which bind variables.

TETk is enriched by new concepts, specific of **TSCL**. One of the most important of them is that of a *well-formed expression* of $\mathscr{L}$.

Sec. II.5. Well-formed expressions

We begin with formulating several conventions, to be used on many occasions further in the text.

C o n v e n t i o n II.3

a. We restrict the range of the variables

(v_1) $p, q, r, s,$ with or without subscripts,

to the set E.

b. The letters

(v_3) $a,b,c,d,\xi,\sigma,$ with or without subscripts,

range over the set I.

c. $\iota(p) = a,\ \iota(q) = b,\ \iota(r) = c,\ \iota(s) = d,$

and for any $i \in N\cup\{0\}$

$$\iota(p_i) = a_i,\ \iota(q_i) = b'_i,\ \iota(r_i) = c_i,\ \iota(s_i) = d_i.$$

d. We enlarge the range of the variables

$i,j,k,l,m,n,\ldots,$ with or without subscripts,

(see Convention I.3) to the set $N\cup\{0\}$ of all natural numbers with zero.

Convention II.3a is not in contradiction with Remark II.1 concerning the variables of the type (v_1) (see Sec.II.2) because

Conclusion II.5c

$$E \subseteq W.$$

The set of all expression tokens is included in the set of all word tokens.

This follows from Conclusions II.5b and II.1a.

The correctness of Convention II.3c follows from Conventions II.3a,b and Conclusion II.5b.

We now formulate the definition of the *set S of all well-formed expression tokens* of a given simple categorial language $\mathcal{L}$. In doing so we avail ourselves of the inductive definition of the set ^nS consisting of all *well-formed expression tokens of order n* ($n\geq 0$).

DEFINITION II.4

a. $$ {}^{0}S = E_{s}, $$

b. $$ \rho \in {}^{k+1}S \Leftrightarrow \rho \in {}^{k}S \vee \bigvee_{n\geq 1} \bigvee_{\rho_0,\rho_1,\ldots,\rho_n \in {}^{k}S} (\rho = \rho(\rho_0,\rho_1,\ldots,\rho_n) \wedge c^{n+1}(\alpha,\alpha_1,\ldots,\alpha_n,\alpha_0)). $$

c. $$ S = \bigcup_{n=0}^{\infty} {}^{n}S. $$

By Definition II.4a a simple expression of $\mathcal{L}$ is a well-formed expression of order 0. By Definition II.4b every well-formed expression token ρ of order $k+1$ either is of order k, and thus is an element of the set ${}^{k}S$, about which we assume that is has already been defined, or is a compound expression of $\mathcal{L}$ with an expansion $\rho_0,\rho_1,\rho_2,\ldots,\rho_n$ such that all its words are well-formed expression tokens of order k, that is expressions which belong to the set ${}^{k}S$. Further, the expression ρ satisfies the condition

$\mathfrak{m}_{\rho}$ $$ c^{n+1}(\alpha,\alpha_1,\ldots,\alpha_n,\alpha_0), $$

in accordance with which the categorial index of the main functor of ρ, that is the α-forming functor, is a concatenation of the index α and the indices of all successive arguments of that functor; in $\mathfrak{m}_{\rho}$ the index α should be followed by a definite technical index. Note in this connection that in Ajdukiewicz's symbolism the categorial index of the main functor of ρ, that is the α-forming index, has the form of the fraction $\alpha/\alpha_1,\alpha_2,\ldots,\alpha_n$.

By Definition II.4c, S is the sum of all sets of well-formed expression tokens of a finite order (greater than or equal to zero).

Instead of speaking about well-formed expression tokens we shall henceforth refer just to well-formed expressions.

The set S of all well-formed expressions of $\mathcal{L}$ we shall - following Ajdukiewicz in that respect - call the *set of syntactically connected expressions.*

Now S can be defined otherwise. To substantiate this claim we adopt a successive definition, which is auxiliary in character.

DEFINITION II.5

$$\delta(X) \Leftrightarrow E_s \subseteq X \wedge \bigwedge_{p} \bigwedge_{n\geq 1} \bigwedge_{p_0, p_1, \ldots, p_n \in X} (p = \rho(p_0, p_1, \ldots, p_n) \wedge$$

$$\wedge\ c^{n+1}(\alpha, \alpha_1, \ldots, \alpha_n, \alpha_0) \Rightarrow p \in X).$$

The set X satisfies the condition $\delta(X)$ if and only if it is a superset of the set E_s and every compound expression with an expansion whose all words are in X and which satisfies the condition $\mathfrak{m}_p$ is also in that set.

We also give the theorem which could replace the adopted definition of S.

THEOREM II.2

$$S = \bigcap \{ X \mid \delta(X) \}.$$

S is the least set that satisfies the condition $\delta(X)$.

The proof is given in the Annex. In that proof we avail ourselves of

LEMMA II.1

$$m \leq n \Leftrightarrow {}^{m}S \subseteq {}^{n}S.$$

The set of all well-formed expressions of an order not greater than n is included in the set of all well-formed expressions of order n.

The lemma follows easily from the fact that , by Definition II.4b, ${}^{k}S \subseteq {}^{k+1}S$ for any k.

Definitions II.4a, II.3c and II.4b, Conclusion II.5b, Axiom II.9 and Definitions II.3b,c yield by induction

LEMMA II.2

$${}^{k}S \subseteq E \subseteq D(\iota),$$

from which, by Definition II.4c, we arrive at

CONCLUSION II.6a

$$S \subseteq E \wedge S \subseteq D(\iota).$$

The set of all well-formed expressions of $\mathcal{L}$ is included in the set of all its expressions and the set of all its words that have an index.

Let us call the elements of $S \setminus {}^{0}S$ *compound well-formed expressions* of $\mathcal{L}$. Definitions II.4c,a,b, Lemma II.2, Axiom II.9, Definition II.3b, and Conclusion II.5a yield

CONCLUSION II.6b

$$S \setminus {}^{0}S \subseteq E_{c} \wedge S \setminus {}^{0}S \subseteq D(\iota) \setminus V.$$

The set of all compound well-formed expressions of $\mathcal{L}$ is a subset of the set of all its compound expressions and the set of all compound word tokens that have a categorial index.

A theorem analogous to Conclusion II.4c holds:

THEOREM II.3

$$p \in S \wedge q \approx p \Rightarrow q \in S.$$

An expression of $\mathcal{L}$ which is equiform with its well-formed expression is also a well-formed expression.

Theorem II.3 follows immediately from the lemma whose proof by induction is in the Annex:

LEMMA II.3

$$p \in {}^{k}S \wedge q \approx p \Rightarrow q \in {}^{k}S.$$

An expression of $\mathcal{L}$ which is equiform with its expression of order k is also a well-formed expression of order k.

From Conclusions II.6a and II.2b and Convention II.3c we arrive at

CONCLUSION II.7

$$p,q \in S \wedge p \approx q \Rightarrow a \approx b.$$

Equiform well-formed expressions of $\mathcal{L}$ have equiform indices.

By Conclusion II.6a, the set S is a set of expressions that have a categorial index. Note that we arrive at

CONCLUSION II.8

$$S \cap \iota(S) = \emptyset.$$

The set of all well-formed expressions of $\mathcal{L}$ has no elements in common with the set of all the indices of these expressions.

Conclusion II.8 is, as it were, a specification of Theorem II.1 and follows from the latter, Conclusions II.6a, II.5c, and Axiom II.7.

The set $\iota(S)$ might be termed the *set of expressions indexing* the well-formed expressions of $\mathcal{L}$ (see the comments that follow Axiom II.6). By Conclusion II.8, the indexing expressions are not included in the expressions of $\mathcal{L}$. They belong to a fragment of a metalanguage of the object language $\mathcal{L}$.

In **TSCL** we adopt one more axiom:

AXIOM II.11

$$\iota(S\setminus{}^{0}S) \cap I_{o} \neq \emptyset.$$

There is a compound well-formed expression of $\mathcal{L}$ which has a basic categorial index.

Let us say that the set I_o of basic indices consists exclusively of indices which are equiform with the letter "s". Those well-formed expressions of $\mathcal{L}$ which have an index assigned to them that is equiform with that letter, are usually called *sentences* or *sentential expressions* of $\mathcal{L}$. In the case now under consideration Axiom II.11 postulates the existence of a sentence which is a compound expression of $\mathcal{L}$.

Axiom II.11 leads us self-evidently to

CONCLUSION II.9a

$$S\setminus{}^{0}S \neq \emptyset.$$

There is a compound well-formed expression of $\mathcal{L}$.

Conclusion II.9a yields immediately

CONCLUSION II.9b

$$S \neq \emptyset.$$

The set of all well-formed expressions of $\mathcal{L}$ is non-empty.

Further, should ${}^{0}S=\emptyset$, then, by Definition II.4b, each of the sets ${}^{k}S$, where $k > 0$, would be empty, too. Hence, by Definition II.4c, S would also be empty, which would contradict Conclusion II.9b. Thus in accordance with Definitions II.4a and II.3a we arrive at

CONCLUSION II.9c

$$^{0}S = E_{S} = D(\iota) \cap V \neq \emptyset.$$

The set of words of $\mathcal{L}$, that is the set of simple word tokens that have an index, is non-empty.

Conclusions II.9a,b and II.6a,b yield the successive

CONCLUSION II.9d. *The sets $D(\iota)$, $D(\iota)\setminus V$, E, and E_C are non-empty.*

We shall now adopt definitions of two sets: the *set B of all basic expressions* of $\mathcal{L}$ and the *set F of all functors* of that language.

DEFINITION II.6

$$B = \{ p \in S \mid \alpha \in I_0 \}.$$

The set of all basic expressions is the set of all those well-formed expressions of $\mathcal{L}$ which have a basic index.

DEFINITION II.7

$$F = \{ p \in S \mid \alpha \in I\setminus I_0 \}.$$

The set of functors is the set of well-formed expressions that have an index which is not a basic index.

It follows from the definitions of B and F and from Axioms II.4 and II.3 that the index of a functor is always a concatenation of (at least two - cf. Theorem I.13^{d}b) indices, whereas the index of a basic expression never is a concatenation of indices. This shows that *B* and *F* are disjoint sets. They are also non-empty. The non-emptiness of *B* follows immediately from Definition II.6 and Axiom II.11 because we have

CONCLUSION II.10

$$(S\setminus{}^{0}S) \cap B \neq \emptyset.$$

There is a compound well-formed expression of $\mathcal{L}$ which is a basic expression.

The non-emptiness of *F* follows immediately from Conclusion II.9a, Definitions II.4c,a,b, Convention II.3b, Theorem I.13^{d}b, and Definition II.7.

It also follows immediately from Definitions II.6 and II.7, Conclusion II.6a, Axiom II.7, and Conclusion I.3^{d} that the sum of *B* and *F* equals the whole set *S*. Hence we have

THEOREM II.4

$$S = B \cup F \wedge B \neq \emptyset \wedge F \neq \emptyset \wedge B \cap F = \emptyset.$$

The set of all well-formed expressions of $\mathcal{L}$ is the sum of two non-empty and disjoint sets: that of all basic expressions and that of all functors.

Since, by Conclusion II.9a, $S\setminus{}^{0}S$ is non-empty, each of the sets ${}^{k}S\setminus{}^{0}S$, where $k > 0$, is non-empty, too, which can be demonstrated on the strength of Definitions II.4c,b,a and Lemma II.1. This is why the set ${}^{1}S\setminus{}^{0}S$, which is equal to the set

$$\{ p \mid \bigvee_{n\geq 1} \bigvee_{p_0, p_1, \ldots, p_n \in {}^0S} (p = \rho(p_0, p_1, \ldots, p_n) \wedge \wedge\, c^{n+1}(\alpha, \alpha_1, \ldots, \alpha_n, \alpha_0)) \}$$

cannot be empty.

Hence there is a $p_0 \in {}^0S$ such that α_0 is a concatenation of a finite number of indices. Hence, by Definition II.4c and Theorem I.13[d]b, there is a $p_0 \in {}^0S$ such that $p_0 \in S$ and $\alpha_0 \in I \setminus I_o$, so that, by Definition II.7, p_0 is a functor. This yields

CONCLUSION II.11

$${}^0S \cap F \neq \emptyset.$$

There is in $\mathcal{L}$ a simple expression which is a functor.

It follows from Conclusion II.10 that we shall always find a basic expression among the compound expressions of $\mathcal{L}$; it follows from Conclusion II.11 and Definitions II.4a and II.3a that we shall always find a functor among the elements of the vocabulary of $\mathcal{L}$.

The following theorems have close connections with Theorem II.3:

THEOREM II.5a

$$p \in B \wedge q \approx p \Rightarrow q \in B.$$

An expression of $\mathcal{L}$ which is equiform with its basic expression is also a basic expression.

and

THEOREM II.5b

$$p \in F \wedge q \approx p \Rightarrow q \in F.$$

An expression of $\mathcal{L}$ which is equiform with its functor is also a functor.

Theorem II.5a follows from Definition II.6, Theorem II.3, Conclusion II.7 and Axiom II.2; Theorem II.5b follows from Definition II.7, Theorem II.3, Conclusion II.7, Theorem I.4^{d}a, and Axiom I.7.

Sec. II.6. Syntactic categories

The concept of syntactic category originated from E.Husserl's *Bedeutungskategorien* (Semantic categories) [1900/01]. Husserl distinguished his semantic categories from the point of view of syntax, namely the role they can play in a sentence, and used them with reference to classes of mutually interchangeable (replaceable) expressions in sentential contexts.[(12)]

A similar intuitive sense was associated with the concept of semantic category by S.Leśniewski, the author of the theory of syntactic categories [1929], A.Tarski [1933], and K.Ajdukiewicz [1935]. In the course of time the more adequate term, namely that of "syntactic category" was adopted by I.M.Bocheński [1949], Y.Bar-Hillel [1950,1964], and contemporary logicians.

The latter do not always use that term in the manner which agrees with its origin.[(13)] In the present book, too, when formulating the definition of that term, we deviate from its usual intuitive sense, which is due to certain difficulties in defining that term, outlined in Sec.II.1.

A *syntactic category with an index* ξ will be symbolized by Ct_{ξ}. We observe Conventions II.3a-c and adopt the following definition of that concept:(14)

DEFINITION II.8

$$Ct_{\xi} = \{ p \mid \alpha \approx \xi \}.$$

A syntactic category with an index ξ *is the set of those expressions of* $\mathcal{L}$ *whose index is equiform with* ξ.

The intuitions connected with the concept of syntactic category defined in this way are explained by the two **fttsc** given in Sec.II.7.

Now Ct will be used to denote the *family of all syntactic categories of expressions of* $\mathcal{L}$, in accordance with the following

DEFINITION II.9

$$Ct = \{ Ct_{\xi} \mid \xi \in \iota(E) \}.$$

The family of all syntactic categories of expressions of $\mathcal{L}$ *is the family of all syntactic categories whose indices are indices of expressions of* $\mathcal{L}$.

By Conclusion II.9d the set E is non-empty. We shall demonstrate that Ct is its logical partition. The relation which determines that partition is the *relation* $\underset{c}{=}$ *of syntactic categorial agreement*. The definition of the relation $\underset{c}{=}$ has the following form:

DEFINITION II.10

$$p \underset{c}{\equiv} q \Leftrightarrow \bigvee_{\xi} p,q \in Ct_{\xi}.$$

The expression p $\underset{c}{\equiv}$ q is read: p and q belong to the same syntactic category.

In accordance with Definition II.10 two expressions belong to the same syntactic category if and only if they belong to a syntacic category with a certain index.

Definitions II.10 and II.8, Conventions II.3a-c, and Axioms I.1a-c yield

CONCLUSION II.12

$$p \underset{c}{\equiv} q \Leftrightarrow a \approx b.$$

Two expressions belong to the same syntactic category if and only if they have equiform indices.

Conclusion II.12 and Axioms I.1a-c yield

CONCLUSION II.13

a. $p \underset{c}{\equiv} p,$

b. $p \underset{c}{\equiv} q \Rightarrow q \underset{c}{\equiv} p,$

c. $p \underset{c}{\equiv} q \wedge q \underset{c}{\equiv} r \Rightarrow p \underset{c}{\equiv} r.$

The relation of syntactic categorial agreement is reflexive, symmetric and transitive in the set E.

In accordance with Conclusion II.13 the relation $\underset{c}{\equiv}$ is an equivalence relation in E and in each of its subsets. It determines the partition of E into classes of abstraction.

It can easily be seen that Conclusion II.12 and Definition

II.8 yield

CONCLUSION II.14

$$[\rho]_{\underset{c}{\equiv}} = Ct_{\alpha}.$$

A class of abstraction with respect to the relation $\underset{c}{\equiv}$ *is a syntactic category with has the index of the expression that is a representative of that class.*

Thus Definition II.9 and Conclusion II.14 yield the previously announced

THEOREM II.6. *The family Ct of all syntactic categories, which is equal to the quotient family* $E/_{\underset{c}{\equiv}}$, *is a logical classification (logical partition) of the set E determined by the relation* $\underset{c}{\equiv}$, *which is to say that Ct is a family of syntactic categories which satisfies the following conditions:*

(i) $\bigwedge_{Ct_\xi \in Ct} Ct_\xi \neq \emptyset$ — *it consists of non-empty categories*

(ii) $\bigwedge_{Ct_\xi, Ct_\sigma \in Ct} (Ct_\xi \neq Ct_\sigma \Rightarrow Ct_\xi \cap Ct_\sigma = \emptyset)$

— *which are pairwise disjoint*

(iii) $\bigcup_{Ct_\xi \in Ct} Ct_\xi = E$ — *and whose sum equals E.*

Let the symbols $Ct(S)$, $Ct(B)$, $Ct(F)$ denote, respectively, the following subfamilies of Ct: the family of all syntactic categories of well-formed expressions of $\mathcal{L}$, the family of all basic categories, and the family of all functor categories, defined as follows:

DEFINITION II.11

a. $Ct(S) = \{ Ct_\xi \mid \xi \in \iota(S) \}$,

b. $Ct(B) = \{ Ct_\xi \mid \xi \in \iota(B) \}$,

c. $Ct(F) = \{ Ct_\xi \mid \xi \in \iota(F) \}$.

It can easily be seen that Definitions II.11, II.6 and II.7, Theorem II.4, Axiom I.1a and Definition II.8 yield

CONCLUSION II.15

$$Ct(S) = Ct(B) \cup Ct(F) \wedge Ct(B) \neq \emptyset \wedge Ct(F) \neq \emptyset \wedge Ct(B) \cap Ct(F) = \emptyset.$$

The family of all syntactic categories of well-formed expressions is the sum of two non-empty and disjoint families of syntactic categories: that of all basic categories and that of all functor categories.

The sets S, B, F are non-empty subsets of E (see Conclusions II.9b and II.6a and Theorem II.4). It follows from Conclusions II.13 that the relation $\equiv_c$ is an equivalence in each of those sets. By Definition II.11 and Conclusions II.14 and II.15 we easily obtain the theorem which explains why the language $\mathscr{L}$ under consideration, described by the system $(\mathscr{L}_c)$ and Axioms II.1–II.11, has been called a *syntactically categorial language*.

THEOREM II.6a. *The relation* $\underset{c}{\equiv}$ *determines a logical classification (logical partition) Ct(S) of S into non-empty, pairwise disjoint syntactic categories whose sum equals S. This is to say that the following conditions are satisfied:*

(i) $\bigwedge_{Ct_\xi \in Ct(S)} Ct_\xi \neq \emptyset,$

(ii) $\bigwedge_{Ct_\xi, Ct_\sigma \in Ct(S)} (Ct_\xi \neq Ct_\sigma \Rightarrow Ct_\xi \cap Ct_\sigma = \emptyset),$

(iii) $\bigcup_{Ct_\xi \in Ct(S)} Ct_\xi = S.$

Further, S can be presented as the sum consisting of the sum of all basic categories which yield B and the sum of all functor categories which yield F, so that the following condition is satisfied:

(iv) $$S = \bigcup_{Ct_\xi \in Ct(B)} Ct_\xi \cup \bigcup_{Ct_\xi \in Ct(F)} Ct_\xi \wedge$$
$$\wedge \bigcup_{Ct_\xi \in Ct(B)} Ct_\xi = B \wedge \bigcup_{Ct_\xi \in Ct(F)} Ct_\xi = F.$$

Sec. II.7. The fundamental theorems of the theory of syntactic categories

As has been mentioned in Sec.II.1 the discussion of **TSCL** results in two theorems. Their formulation will be preceded by several definitions, which are not essential for the formulation of **fttsc** but conclusions drawn from them enable us to better comprehend the nature of both theorems.

We begin with the definition of the *set* C_p^n *of all constituent of n-th order of an expression p* and the definition of the *set* C_p *of all constituents of the expression* p.[15]

The expressions $q \in C^n_p$ and $q \in C_p$ are read, respectively: q is an element of n-th order of p, q is an element of p.

The definition of C^n_p is inductive.

DEFINITION II.12

a. $t \in C^0_p \Leftrightarrow p \approx t.$

A constituent of order zero of a given expression is equiform with that expression.

b. $t \in C^1_p \Leftrightarrow \bigvee_{n \geq 1} \bigvee_{t_0, t_1, \ldots, t_n \in D(\iota)} (p = \rho(t_0, t_1, \ldots, t_n) \wedge$

$\wedge \bigvee_{0 \leq j \leq n} t \approx t_j).$

t is a constituent of first order of a given expression p if and only if p is a compound expression and t is a word which is equiform with some word of the expansion of p.

c. $k > 0 \Rightarrow (t \in C^{k+1}_p \Leftrightarrow \bigvee_{r \in C^k_p} t \in C^1_r).$

A constituent of $k+1$-th order of p, when $k>0$, is a constituent of first order of a constituent of k-th order of p.

d. $t \in C_p \Leftrightarrow \bigvee_n t \in C^n_p.$

A constituent of a given expression is a constituent of a finite order of that expression.

The definitions adopted and Axiom II.10 easily lead us to

CONCLUSION II.16

a. $t \in C_p \wedge q \approx p \Rightarrow t \in C_q,$

b. $t \in C_p \wedge u \approx t \Rightarrow u \in C_p.$

From Definition II.12, Convention II.3a and Conclusions II.5b and II.2a we arrive, by applying mathematical induction, at

CONCLUSION II.17a

$$C_p^n \subseteq C_p \subseteq D(\iota),$$

in accordance with which any constituents of an expression of $\mathcal{L}$ are words of $\mathcal{L}$ which have a categorial index.

If p is a compound well-formed expression, then p does not belong to the set E_s equal ${}^0S$, and the constituents of p which belong to its unambiguously determined expansion (see Axiom II.9) are well-formed expressions by Definitions II.4c,a,b. By applying induction and by referring to Definition II.12 and Theorem II.3 we easily arrive at

CONCLUSION II.17b

$$p \in S \Rightarrow C_p^n \subseteq C_p \subseteq S,$$

in accordance with which any constituents of well-formed expressions are well-formed expressions.

Every constituent of a compound expression occupies in it some syntactic position.[16] Further, to every constituent of a given compound expression we can assign[17] a finite ordered sequence of natural numbers greater than or equal to zero which makes it possible to determine its syntactic position in that expression.

Let the symbol $C_p^{(j_1,j_2,\ldots,j_n)}$ $(n>0)$ denote the *set of the constituents of the expression* p (of order other than 0) *which occupy in it the syntactic position* $(j_1, j_2, \ldots, j_n)$.[18] We adopt the following definition of this term:

DEFINITION II.13

a. $t \in C_p^{(j_1)} \Leftrightarrow p \in E_C \wedge t$ *is equiform with the* j_1*-th word of the expansion of* p,

b. $k > 0 \Rightarrow (t \in C_p^{(j_1, j_2, \ldots, j_{k+1})} \Leftrightarrow t$ *is equiform with the* j_{k+1}*-th word of the expansion of a constituent of the set* $C_p^{(j_1, j_2, \ldots, j_k)}$).

From Definitions II.13a and b and II.12b and c we easily arrive by induction at the following conclusions:

CONCLUSION II.18

$$n > 0 \Rightarrow C_p^{(j_1, j_2, \ldots, j_n)} \subseteq C_p^n.$$

The constituent of p which occupies in it the position determined by the n-term sequence of natural numbers (n>0) is the constituent of n-th order of p.

CONCLUSION II.19

$$n > 0 \Rightarrow (t \in C_p^n \Rightarrow \bigvee_{(j_1, j_2, \ldots, j_n)} t \in C_p^{(j_1, j_2, \ldots, j_n)}).$$

Every constituent of n-th order (n>0) of p occupies in it the syntactic position determined by an n-term sequence of natural numbers.

We now proceed to formulate the definition, important for further analyses, of the four-argument *relation* (/) *of replacement of a constituent of an expression token.* Its formulation requires the adoption, by induction, of the *relation* $(/)^n$ *of the replacement of a constituent of n-th order* $(n>0)$ *of a given expression token.*

The expression $r(p/q)^n s$ is read: the expression r is obtained from the expression s by the replacement of its constituent q of n-th order by the expression p.

The expression $r(p/q)s$ is read analogically as $r(p/q)^n s$, but with the omission of the phrase "of n-th order".

Here are the definitions of the relations under consideration:

DEFINITION II.14

a. $$r(p/q)^0 s \Leftrightarrow s,r \in S \wedge q \approx s \wedge p \approx r.$$

The expression r is obtained from the expression s by the replacement of its constituent q of zero order by the expression p if and only if s and r are well-formed expressions such that q is equiform with s and p is equiform with r.

b. $$r(p/q)^1 s \Leftrightarrow s,r \in S \wedge$$

$$\wedge \bigvee_{n\geq 1} \bigvee_{\substack{s_0,s_1,\ldots,s_n \in S \\ r_0,r_1,\ldots,r_n \in S}} (s=\rho(s_0,s_1,\ldots,s_n) \wedge r=\rho(r_0,r_1,\ldots,r_n) \wedge$$

$$\wedge \bigvee_{0\leq j\leq n} (q \approx s_j \wedge p \approx r_j) \wedge \bigwedge_{\substack{k\neq j \\ 0\leq k\leq n}} s_k \approx r_k)).$$

The expression r is obtained from the expression s by the replacement of its constituent q of first order by the expression p if and only if s and r are compound well-formed expressions with expansions consisting of the same number of well-formed expressions, while in the expansion of s there is an expression equiform with q and having the number j which in the expansion of r determines the expression equiform with p; all expressions in the expansions of s and r which have the same numbers other than j are equiform.

c. $k > 0 \Rightarrow (r(p/q)^{k+1}s \Leftrightarrow \bigvee_{t,v} (r(v/t)^{k}s \wedge v(p/q)^{1}t))$.

The expression r is obtained from the expression s by the replacement of its constituent q of k+1-th order (k>0) by p if and only if r is obtained from s by the replacement of its constituent t of k-th order by an expression v which is obtained from t by the replacement in it of its constituent q of first order by p.

d. $r(p/q)s \Leftrightarrow \bigvee_{n} r(p/q)^{n}s$.

The expression r is obtained from the expression s by the replacement of its constituent q by p if and only if r is obtained from s by the replacement of its constituent q of a finite order (greater than or equal to 0) by p.

The above readings of the labels $r(p/q)^{n}s$ and $r(p/q)s$ has its justification in the conclusions which follow directly from Definitions II.14 and II.12 and Conclusion II.6a:

CONCLUSION II.20

a. $r(p/q)^{n}s \Rightarrow q \in C_{s}^{n} \wedge p \in C_{r}^{n}$,

b. $r(p/q)s \Rightarrow q \in C_{s} \wedge p \in C_{r}$.

In accordance with these conclusions the expression which is replaced in a given expression is its constituent (its constituent of definite order), and the replacing expression is a constituent (its constituent of a definite order) of the expression that is obtained by replacement. Further, the replaced and the replacing constituents are constituents of the same order. Moreover, they are constituents which in compound expressions occupy the same syntactic position because Definitions II.14b and c, II.13a and b yield by induction

Conclusion II.21

$$r(p/q)^n s \wedge n > 0 \Rightarrow \bigvee_{(j_1,j_2,\ldots,j_n)} (q \in C_s^{(j_1,j_2,\ldots,j_n)} \wedge$$

$$\wedge\ p \in C_r^{(j_1,j_2,\ldots,j_n)}).$$

When $r(p/q)s$, p and q are often called *replaceable* or *interchangeable* in s and r. Such terminology has its justification in the following conclusion that is obtained easily from Definition II.14 and Axiom I.1b:

Conclusion II.22

$$r(p/q)s \Rightarrow s(q/p)r.$$

If r is obtained from s by the replacement of its constituent q by p, then s is obtained from r by the replacement of its constituent p by q.

Those constituents which are replaceable in expressions are likewise well-formed expressions because we have

Conclusion II.23

$$r(p/q)s \Rightarrow s,r,q,p \in S,$$

which follows from the fact that if $r(p/q)s$, then s and r are well-formed by Definition II.14, and their constituents p and q (cf. Conclusion II.20) are well-formed, too, in view of Conclusion II.17b.

Once we have described certain properties, which are in agreement with our intuitions, connected with the replacement of expressions we formulate and prove the first of the two **fttsc**.

THEOREM II.7a (I **fttsc**)

$$r(p/q)s \wedge r \underset{c}{=} s \Rightarrow p \underset{c}{=} q.$$

If an expression r, which belongs to the same syntactic category as an expression s, is obtained from the latter by the replacement of its constituent q by an expression p, then p and q belong to the same syntactic category.

P r o o f. It suffices to prove, by applying the principle of induction, that the following lemma holds:

LEMMA II.4

$$r(p/q)^n s \wedge r \underset{c}{=} s \Rightarrow p \underset{c}{=} q.$$

I **fttsc** follows immediately from that lemma by Definition II.14d.

Now when $n=0$ Lemma II.4 is obtained directly from Definition II.14a, Theorem II.3, Conclusions II.7, II.12, II.13b,c, and Conventions II.3a,c.

Assume that

(1a) $$r(p/q)^1 s$$

and

(1b) $$r \underset{c}{=} s.$$

We shall prove that $p \underset{c}{=} q$ and hence that Lemma II.4 holds for n=1.

We conclude from the assumption (1a) and Definition II.14b that

(2) $$r, s \in S$$

and that there are expressions $s_0, s_1, \ldots, s_{n_1} \in S$ $(n_1 \geq 1)$ and $r_0, r_1, \ldots, r_{n_1} \in S$ such that

(3) $$s = \rho(s_0, s_1, \ldots, s_{n_1})$$

and

(4) $$r = \rho(r_0, r_1, \ldots, r_{n_1})$$

and that for a certain natural number j_1 such that

(5) $$0 \leq j_1 \leq n$$

we have

(6) $$q \approx s_{j_1}$$

and

(7) $$p \approx r_{j_1}$$

and also

(8) $$\bigwedge_{0 \leq k \leq n_1} (k \neq j_1 \Rightarrow s_k \approx r_k).$$

Since s and r are well-formed and compound expressions - by formulas (2)-(4), Conclusion II.6a, Definition II.3b, and Axiom II.9 - they belong to the set $S \setminus {}^0S$ (see Definitions II.3a and II.4a) and are at the same time expressions of an order greater than zero (Definition II.4c). By Axiom II.9 the function ρ is an injection. Hence the expansions of s and r are determined unambiguously, and in view of Definition II.4b and the formulas

(3) and (4) s and r satisfy the conditions $\mathfrak{m}_s$ and $\mathfrak{m}_r$, respectively. Under Convention II.3c we record these conditions, respectively, in the form

(9) $$c^{n_1+1}(d,d_1,d_2,\dots,d_{n_1},d_0)$$

and

(10) $$c^{n_1+1}(c,c_1,c_2,\dots,c_{n_1},c_0).$$

Since the constituents of s and r which belong to their expansions are well-formed expressions and the formula (8) holds, in view of Convention II.3c we can state, on the strength of Conclusion II.7, that

(11) $$\bigwedge_{0\le k\le n_1}(k \neq j_1 \Rightarrow d_k \approx c_k).$$

It follows from the assumption (1b), which has not been referred to so far, from Conclusion II.12 and Axiom I.1b that, under Convention II.3, the indices d and c of the expressions s and r, respectively, are equiform so that

(12) $$d \approx c.$$

In accordance with (5) there can be two cases: j_1=0 or $1\le j_1\le n_1$. In the first case, we avail ourselves of Theorem I.17 on generalized concatenation and find, by the formulas (9), (10), (12) and (11), that $d_{j_1} \approx c_{j_1}$ so that - in view of Conclusions II.23, II.7, II.12 and II.13b,c and the formulas (6) and (7) - $p \underset{c}{\equiv} q$. It is likewise in the second case, for $1\le j_1\le n$. This time it suffices to refer to Theorem I.18 on generalized concatenation in order to find, in view of (9), (10), (12) and (11), that $d_{j_1} \approx c_{j_1}$, from which it follows that

$p \underset{c}{=} q$.

We have thus demonstrated that Lemma II.4 holds for $n=1$.

Assume now that this lemma holds for $k>0$ and $n=k$. Assume also that $r(p/q)^{k+1}s$ and $r \underset{c}{=} s$.

We shall demonstrate that under these assumptions $p \underset{c}{=} q$, and hence that it follows from the assumption that Lemma II.4 holds for $n=k$ $(k>0)$ that it also holds for $n=k+1$.

By Definition II.4c it follows from the assumption $r(p/q)^{k+1}s$ that there are expressions v_1 and t_1 such that $r(v_1/t_1)^k s$ and $v_1(p/q)^1 t_1$. From the inductive assumption and the assumption stating that $r \underset{c}{=} s$ we infer that $v_1 \underset{c}{=} t_1$. Since $v_1(p/q)^1 t_1$ and Lemma II.4 has been proved for $n=1$, we have that $p \underset{c}{=} q$.

This concludes the inductive proof of Lemma II.4.

We now proceed to formulate and to prove the second **fttsc**.

THEOREM II.7b (II **fttsc**)

$$r(p/q)s \wedge p \underset{c}{=} q \Rightarrow r \underset{c}{=} s.$$

If an expression r is obtained from an expression s by the replacement of its constituent q by p which belongs to the same syntactic category as q, then r and s belong to the same syntactic category.

P r o o f. The second **fttsc** follows from Definition II.14d and

LEMMA II.5

$$r(p/q)^n s \wedge p \underset{c}{=} q \Rightarrow r \underset{c}{=} s,$$

whose proof, as in the case of Lemma II.4, is by induction with.

respect to the order n of the constituent q of s.

For $n=0$ Lemma II.5 follows self-evidently from Definition II.14a, Theorem II.3, Conclusions II.7, II.12, II.13b,c and Conventions II.3a,c.

Note also that for $n=1$ the assumption

(1a) $$r(p/q)^{1}s$$

yields the formulas (2)-(11) used in the proof of Lemma II.4, and the assumption

(1b) $$p \underset{c}{\equiv} q$$

and the formulas (6) and (7) yield the formula

(13) $$d_{j_1} \approx c_{j_1}$$

(in view of Conclusions II.23, II.7, II.12 and II.13b,c and in accordance with Convention II.3c).

In order to find that Lemma II.5 holds for $n=1$ it suffices to demonstrate that $r \underset{c}{\equiv} s$. In fact, since, by (5), $0 \leq j_1 \leq n_1$ and the formulas (11) and (13) hold, we have

$$\bigwedge_{0 \leq k \leq n_1} d_k \approx c_k.$$

Since the formulas (9) and (10) hold we may refer to Theorem I.18 and conclude that $d \approx c$, so that, by Axiom I.1b, Conclusion II.12 and Convention II.3c we have that $r \underset{c}{\equiv} s$.

It remains to prove that if Lemma II.5 holds for a natural number $n=k$ ($k>0$), it also holds for a natural number $n=k+1$.

Assume accordingly that the inductive assumption holds and that the same applies to the formulas $r(p/q)^{k+1}s$ and $p \underset{c}{\equiv} q$. We shall demonstrate that in such a case $r \underset{c}{\equiv} s$. Since $r(p/q)^{k+1}s$ we can infer, by Definition II.14c, that there are expressions

v_1 and t_1 such that $r(v_1/t_1)^k s$ and $v_1(p/q)^1 t_1$. Since $v_1(p/q)^1 t_1$ holds we immediately have that $v_1 \underset{c}{\equiv} t_1$ because we have assumed that $p \underset{c}{\equiv} q$ and Lemma II.5 has been proved for n=1. Since $r(v_1/t_1)^k s$, $r \underset{c}{\equiv} s$ follows immediately from the inductive assumption.

The two **fttsc** taken jointly make it possible to prove

THEOREM II.7

$$r(p/q)s \Rightarrow (p \underset{c}{\equiv} q \Leftrightarrow r \underset{c}{\equiv} s).$$

Two expressions of $\mathcal{L}$ *belong to the same syntactic category if and only if on replacing one of them by the other in a well--formed expression of* $\mathcal{L}$ *we obtain a well-formed expression of* $\mathcal{L}$ *which belongs to the same syntactic category as the former.*

Theorem II.7 cannot, of course, be used as a definition of belonging of two expressions to the same syntactic category. It confirms the correctness of the definition of syntactic category adopted in Section II.6.

Sec. II.8. The algorithm of checking the syntactic correctness of expressions

The language $\mathcal{L}$, which is the object of investigation by the theory **TSCL**, ahould be *decidable*, which is to say that there should be an algorithm enabling us to check whether a given expression belongs to the set S of the well-formed expressions of $\mathcal{L}$. The algorithm to be described below is a generalized procedure defined in Ajdukiewicz [1935] and intended to investigate the *syntactic connectedness*, i.e. *well-formedness* or *syntactic correctness* of expressions. It also refers to certain ideas to be found in Ajdukiewicz [1960]. That algorithm pertains to compound expressions of $\mathcal{L}$. When formulating further comments about it we shall not always observe the rigour of precision, but there seems to be no point in making those comments more precise.

We begin with several general remarks on the concept of algorithm, which, it seems, are somewhat different from those which are usually made on such an occasion. We do not, of course, claim them to be original.

An *algorithm* is a sequence of psychophysical operations or operations performed by a machine, determined by appropriate directives.[19] Those directives should have several essential properties[20] such as *performability* (each directive should secure the possibility of an effective performance of a given operation), *finiteness* (the number of the directives should be finite), and *definiteness* (one of them should define the first operation, others should indicate the successive operation once the results of the earlier operations are known, and the last directive should inform one about the characteristic feature of the final operation of a given algorithm).

When formulating the directives which determine the algorithm of checking the well-formedness of the expressions of a given language $\mathcal{L}$ we avail ourselves from several assumptions about the symbolism of $\mathcal{L}$. First of all we assume that every label of $\mathcal{L}$ is recorded with a finite number of symbols. Next we assume that about any two labels of $\mathscr{L}$ we can decide whether they are equiform, and about any three labels, whether any of them is the concatenation of the remaining two. We also assume that about any label we can decide whether it is a word from the vocabulary of $\mathscr{L}$, whether it is a word that has an index, and whether it is an expression of $\mathscr{L}$. Further we assume that for any given compound expression of $\mathscr{L}$ we can indicate which parts of it (in the sense of Leśniewski's mereology) correspond to the words of its expansion, which part of it corresponds to its main functor, and which parts of it correspond to the successive arguments of that functor. Finally we assume that for every expression of $\mathscr{L}$ we can indicate the categorial index assigned to it.

In accordance with the assumptions made above the following labels, chosen by way of example, are not subject to checking: the label in the notation of the sentential calculus, $\Rightarrow(p\wedge q)\neg$, and the arithmetical label $x+y\cdot z$, because we lack the additional information about the binding force of the functors $\cdot$ and +. These labels are not compound expressions; we cannot indicate in them the main functors and their arguments. The second label, however, is a word which has an index because it is treated as a structurally ambiguous name expression belonging to arithmetic.

We adopt two terminological conventions.

The sequence of categorial indices

$$(\xi) \qquad \xi_0, \xi_1, \xi_2, \ldots, \xi_n$$

is called the *intermediate expansion of the expression*

(δ) $\rho(p_0, p_1, \ldots, p_n)$

which has its expansion in the sequence of expressions

(ε) $p_0, p_1, \ldots, p_n$

if and only if for any i $(0 \leq i \leq n)$ ξ_i is equiform with the i-th term of the sequence

(ζ) $\alpha_0, \alpha_1, \ldots, \alpha_n$.

Note that in accordance with Conventions II.3b,c every i-th $(0 \leq i \leq n)$ term of the sequence (ζ) is the categorial index of the i-th term of the sequence (ε).

It is said about the sequence (ξ) that it *satisfies the condition* $\mathfrak{m}_p$ if and only if the index ξ_0 is the concatenation of the index α of the expression p and of the indices $\xi_1, \xi_2, \ldots, \xi_n$, so that the following holds:

$\mathfrak{m}_p$ $c^{n+1}(\alpha, \xi_1, \xi_2, \ldots, \xi_n, \xi_0)$.

It is assumed about every finite sequence of categorial indices that it is possible to decide whether it does, or does not, satisfy the condition $\mathfrak{m}_p$.

Note that if the sequence (ξ) is an intermediate expansion of the expression (δ) then the latter sequence satisfies the condition $\mathfrak{m}_{(\delta)}$ if and only if that condition is satisfied by (ζ) (see Condition I.11).

Before we formulate the directives which determine the algorithm for checking the well-formedness of expressions, we shall describe an auxiliary algorithm, namely that of drawing the *tree of derivation* (also termed the dendrite) of a given compound expression token. Such a diagram will henceforth be termed tree I of a given expression.[21]

By rewriting the compound expression whose well-formedness we want to check we obtain the first (top) line of tree I. By writing immediately below, on one and the same level, from left to write, the words equiform with the successive components of the expansion of the expression under consideration we obtain the second line of tree I. If every such word is a simple expression or if some of them is a compound word which is not a compound expressions (and hence is a word that cannot be decomposed into its main functor and its arguments), then the tree I of the expression in question has been completed. If, on the contrary, not every such word is a simple expression, and if compound words of the expansion of the given expression are compound expressions, then under each such expression we write words which are equiform with the successive words of its expansion, in the same way as in the case of line two. In this way we obtain line three of tree I. We proceed in this way until we obtain a line whose all elements are simple expressions or some of them is a compound words which is not a compound expressions. By writing down the last element of such a line we perform the last operation of the construction of tree I of the compound expression in question. The conclusion that such an element is obtained in a finite number of steps follows from the fact that in each line except the last (lowest) one there is an element which is recorded by means of a greater number of simple words than every element of the next line.

If in tree I of a given expression some of its final nodes

is not a simple expression (is a compound word that is not an expression), then the algorithm of checking the well-formedness of that expression is completed: the expression under consideration is not well-formed (in the terminology of Ajdukiewicz [1935] it is not *thoroughly well arranged*). If all the final nodes of tree I of a given compound expression are simple expressions from the set ${}^0S$ (in Ajdukiewicz's terminology, if the expression under consideration is *thoroughly well arranged*), then we proceed to perform the successive operations of checking the well-formedness of that expression.

The elements of tree I of a thoroughly well-arranged expression are ordered so that out of any two elements which are in different lines that one is the *earlier* which is in the higher line, and out of any two elements in the same line that one is the *earlier* which is situated more to the left.(22)

Once tree I has been completed, we construct tree II, which might be termed the *tree of categorial indices* of a given compound expression. Tree II is obtained from tree I by the successive replacement in it, in accordance with the order established above, of all its elements by their categorial indices.

Before formulating the directives of the principal part of the algorithm of checking the well-formedness of a thoroughly well - arranged compound expression of $\mathcal{L}$ we order the intermediate expansions of compound expressions of tree I, formed of parts of tree II. Of any two such expansions that one is the *earlier* which is an expansion of an earlier element of tree I.

Those parts of tree II which form intermediate expansions of all compound expressions $s_0, s_1, \dots, s_k$ $(k \geq 1)$ of tree I of a given expression are checked so that we find - in accordance with the order fixed for intermediate expansions - whether they do, or do not, satisfy the conditions m_{s_j} $(0 \leq j \leq k)$. Once we encounter the first intermediate expansion out of the such expansions of the expressions s_j $(0 \leq j \leq k)$ which does not satisfy the condition m_{s_j}, the algorithm of the checking the syntactic correctness of expressions has been performed, and we conclude that the expression under consideration is not well-formed. If we find that for any compound expression s_j $(0 \leq j \leq k)$ which is an element of tree I its intermediate expansion formed of a part of tree II satisfies the condition m_{s_j}, then the algorithm of checking the well-formedness of expressions has been performed and we conclude that the expression under consideration is well-formed.

By making use of the algorithm described above we shall check now the syntactic correctness of two expressions from the arithmetic of real numbers and one expression from the sentential calculus. In the recording of those expressions we use brackets as technical symbols which have no indices and are used in the ordinary way.

E x a m p l e II.1. Assume that the variables x and y range over the set of real numbers, and the symbols +, <, and 0 denote, respectively, the operation of addition, the relation "less than", and the number zero. We proceed to check the syntactic correctness of the expression

(π1) $x + (y < 0)$.

We construct tree I of the expression (π1) (see Fig.II.1).

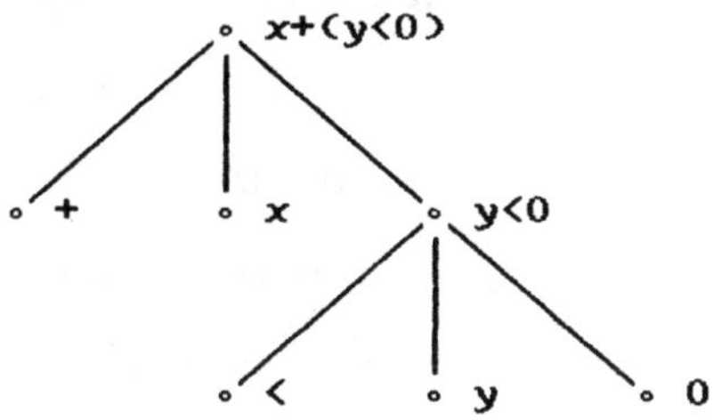

Fig.II.1. Tree I of the expression (π1)

As can be seen, then expression (π1) is thoroughly well-arranged. We accordingly construct tree II of the categorial indices of the expression (π1) (see Fig.II.2).

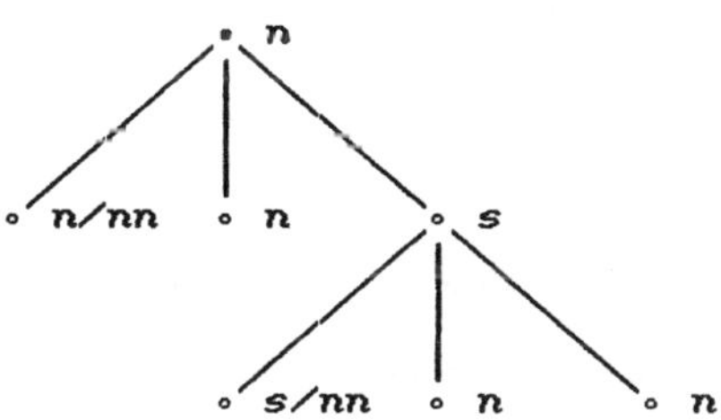

Fig.II.2. Tree II of the expression (π1)

Trees I and II of the expression under consideration consist of three lines each. The lines of tree II form two intermediate expansions: that of the expressions which is equiform with the expression being checked and that of the expression which is equiform with $y<0$. The letters which are equiform with n and s

- and which are parts of tree II - are, respectively, indices of expressions which belong to the category of name expressions, and indices of expressions which belong to the category of sentential expressions. For typographical reasons the indices of functors (concatenations of certain indices) are recorded in the "quasi-fractional linear" notation using symbols which are equiform with / as technical symbols. The index which occurs before the slant / indicates the index of the expression which a given functor forms, and the indices which occur after that symbol indicate, successively, the indices of the successive arguments of that functor. When carrying out the essential operations of the algorithm of checking the well-formedness of expressions we find that in the case of the expression (π1) the first intermediate expansion determined by tree II does not satisfy the required condition. The index of the main functor of the expression which is the root of tree I is not a concatenation of the index of that expression (the root of tree I) and the indices of the successive arguments of that fuunctor. Hence the expression (π1) is not a well-formed arithmetical expression.

E x a m p l e II.2. We check the syntactic correctness of the expression from the sentential calculus:

(π2) $$p \Rightarrow (\neg\ (\neg)\ \vee\ p).$$

Its tree I is shown in Fig.II.3. It can be seen that the expression (π2) is thoroughly well-arranged. Its tree II is shown in Fig.II.4.

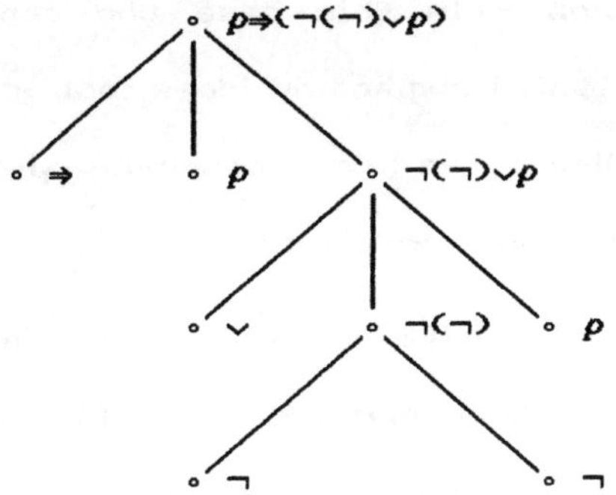

Fig.II.3. Tree I of the expression (π2)

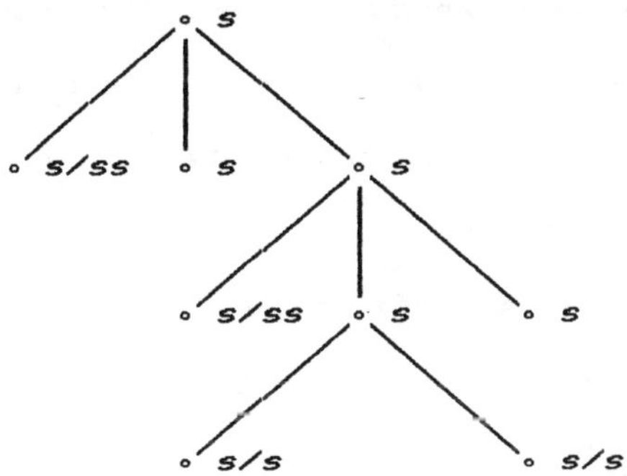

Fig.II.4. Tree II of the expression (π2)

Trees I and II of the sentential calculus expression (π2) now under consideration consist of four lines each. Tree II of that expression determines three intermediate expansions. When carrying out the basic operations of the algorithm - in accordance with the order of those expansions - we find that the first expansion satisfies the desired condition so that the index of the functor of implication is a concatenation of the index of the expression which that functor forms (i.e., the expression (π2)) and the indices of the successive arguments of that functor. The second intermediate expansion also satisfies the desired condition - in this case the condition $\mathfrak{m}_{\neg(\neg)\vee p}$. On the contrary, the third expansion does not satisfy the desired condition $\mathfrak{m}_{\neg(\neg)}$. Hence we can conclude that the expression under consideration is not well-formed.

E x a m p l e II.3. We shall now examine the well-formedness of another arithmetical expression, namely

$$(\pi 3) \qquad x \cdot (y + (-z)) = (x \cdot y) + (x \cdot (-z)).$$

In recording that expression we observe the usual conventions pertaining to the constants $\cdot$, $+$, $-$, the variables x,y,z, and the brackets. We construct tree I and tree II of this expression and note that each tree has five lines and determines eight expansions.

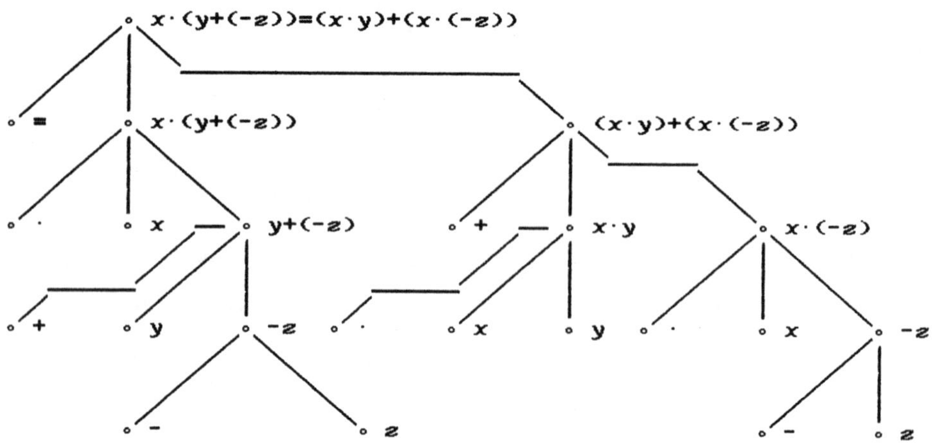

Fig.II.5. Tree I of the expression $(\pi 3)$

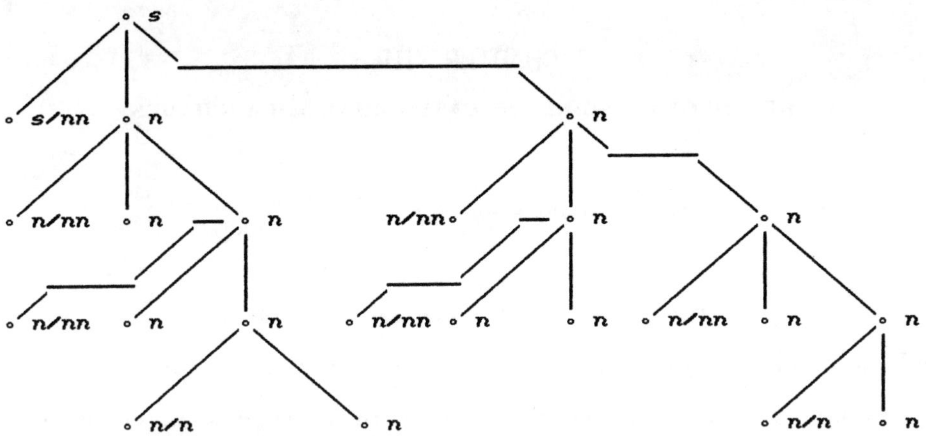

Fig.II.6. Tree II of the expression (π3)

In carrying out the basic operations of the algorithm of checking the syntactic correctness of (π3) we find that each of the eight intermediate expansions determined by tree II of (π3) satisfies the desired condition so that the index of the main functor of each compound expression which is an element of tree I of (π3) is a concatenation of the index of the expression which that functor forms and the indices of the successive arguments of that functor. This shows that the expression under consideration is well-formed.

CHAPTER III

THE THEORY TSCω-L OF CATEGORIAL ω-LANGUAGES

Sec. III.1. Introductory remarks

The theory **TSCL** described languages which are poor in a certain sense. This is because most theorems formulated in mathematical theories are formulated in languages which include new expressions, not occurring in simple categorial languages, namely expressions with variables and operators that bind those variables. It seems, therefore, natural to modify **TSCL** - while retaining the main objectives described in Sec.II.1 and observed when **TSCL** was constructed - so that it should cover the languages mentioned above. The syntactic analysis of expressions, correct from the point of view of the theory of syntactic category, should be respected, too.

In such an analysis we usually make use of categorial indices in accordance with the principle of splitting expressions into "main functors and their arguments".

The method of using indices in the study of the syntactic structure of linguistic expressions which can include variables and operators that bind them has encountered many problems, mentioned already by Ajdukiewicz in [1935]. One of them is the trouble connected with the algorithm of checking the syntactic connectedness of the expressions now under consideration; the other is the "typical ambiguity" of the operators which bind variables belonging to different syntactic categories. Thus, for instance, if we admit the distinction, in the language of the arithmetic of natural numbers, between two different syntactic categories of name variables: that of individual variables and that of variables which stand for the names of sets, then the equiform operators of abstraction in such expressions as, e.g., $\{x \mid x>5\}$ and $\{X \mid X \subseteq N\}$ would have different indices.

It is likewise when we consider expressions such as

$$\bigwedge_{p} (p \vee \neg p) \quad \text{and} \quad \bigwedge_{x} x = x,$$

in which the operators, namely the universal quantifiers, equiform with the symbol $\wedge$, bind, respectively, a variable of the category of sentences and a variable of the category of names.

The elimination of the difficulties related to the "typical ambiguity" of operators may consist in the appropriate marking of a syntactically ambiguous operator and thus splitting it into diversiform signs in accordance with definite caligraphic conventions (cf. R.Suszko [1958], [1964]). For instance the symbols $\overset{0}{\wedge}$ and $\overset{1}{\wedge}$ could be used to denote the universal quantifier which binds a sentential variable and the universal quantifier which binds a name variable, respectively. In the conception of L.Borkowski [1958] the obstacle connected with the typical ambiguity of (proper) quantifiers which bind variables belonging to different syntactic categories could be removed by the adoption of the convention that the index of the syntactic category of quantifiers is independent of the index of the quantifier variable (the variable which accompanies the quantifier).

The problems connected with the indices, and hence with the syntactic categories, of the operators which bind variables could be eliminated in general if the operators could be defined by expressions which do not include quantifiers. Ajdukiewicz [1935] mentioned the possibility of defining the universal quantifier by means of terms in which operators do not occur. In connection with this it is worth-while mentioning the result obtained by J.Słupecki [1955]: every thesis of the full system of the simple type theory can be brought into one-to-one correspondence with a synonymous thesis of a logical system in the language of which bound variables and operators do not occur. This points to the possibility of eliminating quantifiers from the full system of the simple type theory (without the axiom of infinity). As it is known, the full system of the simple type theory enriched with the axiom of infinity suffices to form the foundations of mathematics. Is there, however, a possibility of eliminating quantifiers and other operators from the language of mathematics?

Operators have become deeply rooted in mathematical practice, and it is therefore worth-while seeking a solution which, while retaining the basic ideas of the theory of syntactic categories, would make it possible to approach the problem of the syntactic connectedness of expressions which include operators and the variables bound by them by applying

Ajdukiewicz's method of making use of indices.

It is worth noting that in the Polish literature of the subject endeavours have been made to modify Ajdukiewicz's method so that it could cover the role of variables and the operators that bind them. Those endeavours, however, deviate more or less from the general principles of the theory of syntactic categories. For instance, in the works of R.Suszko [1958], [1960], [1964] indices are used as instruments of co-ordinating expressions and the objects which they denote (see also A.Nowaczyk [1978] and B.Stanosz and A.Nowaczyk [1976]).

In the present chapter we set ourselves the task of modifying **TSCL** in such a way - while retaining Ajdukiewicz's ideas and complying with the goals of constructing a theory of syntactically categorial languages, as discussed in Sec.II.1 - as to be able to discuss the syntactically categorial languages of expression tokens in which operators and the variables bound by them may occur. Such languages will be termed *ω-languages*. To simplify our analysis we adopt the convention that the expressions occurring in those languages may include operators which bind only one variable each and that the range of such operators is confined to one expression.[(23)] It does not seem, however, that the considerations carried out in this chapter could not be generalized so as to cover the languages which include expressions with quantifiers that bind a finite number of variables and the languages which include expressions with operators that have a finite number of arguments.

It is the theory **TSCω-L** which is the formal theoretical conception pertaining to the ω-languages. It is concerned with any fixed ω-language $\mathcal{L}$. That language is characterized by an ordered system

$$(\omega\text{-}\mathcal{L}_c) \quad \langle Lb;\ V,\ Vr,\ O,\ W;\ \approx,\ c;\ I_o, I;\ \iota,\ \rho,\ (fv);\ E,\ S,\ Ct(S)\rangle,$$

whose elements satisfy definite conditions formulated by the axioms and definitions of **TSCω-L**. In $(\omega\text{-}\mathcal{L}_c)$, next to the concepts occurring in the ordered system

$$(E_c) \qquad \langle\ Lb;\ V,\ W;\ \approx,\ c;\ I_o,\ I;\ \iota,\ \rho;\ E\ \rangle,$$

described by the theory **TETk** of expression tokens (see Sec.II.4) and to the concepts of the set S of all well-formed expressions and the family $Ct(S)$ of all syntactic categories of S, which occur also in $(\mathcal{L}_c)$ but are, for obvious reasons, described in **TSCω-L** somewhat differently than in **TSCL**, there are new concepts: the *set* Vr *of all variables*, the *set of operators* O, and the *relation* (fv) *of being a free variable in a given expression token*. They will be discussed in Sec.III.3. Their inclusion in the conceptual apparatus of **TSCω-L** shows that its store of concepts is richer than that of **TSCL**.

Before we present the axioms, basic definitions and theorems of **TSCω-L** we shall point to certain relationships which link that theory to **TLTk** and its expansion, that is the theory **TETk** of expression tokens, which is a fragment and at the same time the core of **TSCL** (Sec.III.2).

The definition of the concept of well-formed expression adopted in **TSCω-L** (Sec.III.4) is a modification of the definition of that concept adopted in **TSCL** (Sec.II.5). It makes it possible to formulate the fundamental theorems which describe the syntactically categorial ω-languages (Sec.III.5) and which are analogous to Theorems II.2, II.4, II.6a and II.7a,b (I and II **fttsc**). Further it makes it possible to describe the algorithm of checking the syntactic connectedness (syntactic correctness) of those expressions which may include operators that bind variables (Sec.III.6).

Sec. III.2. The foundations of TSCω-L

The theory **TSCω-L** is based on the theory **TETk** of expression tokens while the latter is in turn based on the theory **TLTk** of label tokens.

Note (see Sec.II.4) that **TETk** is obtained from **TLTk** by having its apparatus enriched by new primitive and secondary concepts and by axioms and definitions (Axioms II.1–II.10 and Definitions II.1a, II.1, II.2a,b, II.3a–c, II.8–II.10, II.12a–d, II.13a,b) which establish the properties of those concepts.

In connection with this and in accordance with Conventions I.1a,b and II.3b the following remark is to be made:

REMARK III.1. The constant symbols of the vocabulary of **TETk**, namely Lb, V, W, $\approx$, c, I_o, I, ι, ρ, E, Ct, $\underset{c}{=}$, C_p are constant symbols of the vocabulary of **TSCω-L**.

The variable symbols of the vocabulary of **TSCω-L** include the letters

(v_{11}) $p,q,r,s,t,u,v,\ldots$, with or without subscripts,

which range over *Lb*; the letters

(v_{12}) a,b,c,d,ξ,σ, with or without subscripts,

which range over *I*; the letter

(v_2) *X*, with or without subscript,

which ranges over the family 2^{Lb}.

The preliminary remarks and Remark III.1 enable us to formulate two metatheorems pertaining to **TSCω-L**.

THEOREM III.1*. *Every thesis (accepted sentence) of* **TETk** *is a thesis of* **TSCω-L**.

THEOREM III.2*. *Every thesis (accepted sentence) of* **TLTk** *is a thesis of* **TSCω-L**.

Note that we have retained a convention analoguous to Conventions I.2 and II.1, namely

C o n v e n t i o n III.1. The theses (accepted sentences) whose numbers are marked with an asterisk are theses of the metatheory of **TSCω-L**.

Theorem III.1* enables us to arrive at the conclusion to which we shall refer on numerous occasions.

CONCLUSION III.1*. *The theses (accepted sentences) of* **TSCω-L** *include: Axioms II.1–II.10; Definitions II.1a, II.1, II.2a,b, II.3a–c, II.8–II.10, II.12a–d, II.13a,b; Theorem II.1; Conclusions II.1a,b, II.2a,b, II.3, II.4a–c, II.5a–c, II.12–14, II.16a,b, II.17a, II.18, II.19.*

By retaining the definition (see Sec.II.2) of the dual analogon of an expression of **TLTk** we obtain – by Theorem III.2*, Remarks II.3 and II.2 made in Sec.II.2, and Conclusion III.1* – the conclusion which is analogous to Theorem II.3*, namely

CONCLUSION III.2*. *Every dual analogon of a thesis (accepted sentence) in* **TLTk** *is a thesis in* **TSCω-L.**

In connection with the relationships shown above it is convenient, when presenting **TSCω-L** is the successive sections of this chapter, to observe the conventions introduced in Chap.II. Hence we have

C o n v e n t i o n III.2. We retain Convention II.2 (Sec.II.2) and Conventions II.3a-d (Sec.II.5).

Sec.III.3. Operator expressions

We now add to the conceptual apparatus of **TETk** two new concepts which do not occur in **TSCL** and which are primitive concepts in **TSCω-L**. They are: the *set O of all operators* of the ω-language $\mathcal{L}$, and the *set Vr of all variables* of that language. The intuitive sense which we impart to the elements of those sets, that is, respectively, operators and variables, does not deviate from the one which we impart to them in the languages of formal disciplines. Note, however, that the term "operator" will be used to denote a label which does not include a variable. Thus, for instance, by the universal quantifier we mean the graphic sign $\wedge$ and not such graphic symbols as $\underset{x}{\wedge}$, $\underset{z}{\wedge}$, and by the algebraic sign of summation we mean Σ and not, say, $\sum_{k=1}^{\infty}$.

This is so because we adopt

AXIOM III.1

$$O \cup Vr \subseteq E_s.$$

The set of all operators and the set of all variables are subsets of the set of simple expression tokens.

In accordance with Axiom III.1 and Definition II.3a we arrive at

CONCLUSION III.1

a. $O \subseteq V \wedge O \subseteq D(\iota)$,

b. $Vr \subseteq V \wedge Vr \subseteq D(\iota)$.

Thus operators and variables are simple word tokens of the ω-language $\mathscr{L}$ under consideration which have indices.

In the vocabulary V of the ω-language $\mathscr{L}$ there are no operators that would be "typically ambiguous", because Conclusions II.2b and III.1a and Convention II.3c (see Convention III.2) yield

CONCLUSION III.2a

$$p,q \in O \wedge p \approx q \Rightarrow a \approx b.$$

The indices of equiform operators are equiform.

Likewise, Conclusions II.2b and III.1b and Convention II.3c yield

CONCLUSION III.2b

$$p,q \in Vr \wedge p \approx q \Rightarrow a \approx b.$$

The indices of equiform variables are equiform.

We also assume about the sets *O* and *Vr* that they satisfy, respectively, axioms which are analoguous to Axiom I.10:

AXIOM III.2

$$p \in O \wedge t \approx p \Rightarrow t \in O.$$

A label token which is equiform with an operator is also an operator.

AXIOM III.3

$$p \in Vr \wedge t \approx p \Rightarrow t \in Vr.$$

A label token which is equiform with a variable is also a variable.

It is often emphasized that one of the essential functions of the operators is that operators can bind variables. The binding role of an operator is revealed in definite structures which include operators. To describe such structures in the ω-language $\mathcal{L}$ we distinguish from the set *E* of all expression tokens of $\mathcal{L}$ its subset $\rho(O\text{x}Vr\text{x}E)$, whose elements will be termed *operator expression tokens* of the ω-language $\mathcal{L}$ or, briefly, *operator expression* of that language. Operator expressions are thus the values of the function ρ restricted to the set $O\text{x}Vr\text{x}E$. They are thus compound expressions consisting of an operator, a variable, and an expression token, in that order.

In our further analyses we shall make use of the

abbreviation introduced by the metalinguistic

DEFINITION III.1*

$$p = \rho^{\omega}(p_0,p_1,p_2) \underset{\mathrm{df}}{=\!=} p = \rho(p_0,p_1,p_2) \wedge p_0 \in O \wedge p_1 \in Vr \wedge p_2 \in E,$$

which is in fact a rule which allows us to replace, in every thesis of the theory **TSCω-L** now under construction, an expression equiform with that standing on the one side of the sign $\underset{\mathrm{df}}{=\!=}$ by an expression equiform with that standing on the other side of that sign, and to replace an expression equiform with a substitution for the expression standing on the one side of the said sign by an appropriate substitution for the expression standing on the other side of that sign.

The elements of the expansion of the operator expression

$$(\delta^{\omega}) \qquad \rho^{\omega}(p_0,p_1,p_2),$$

i.e., the elements of the sequence p_0,p_1,p_2, are called, successively, p_0 - the *main operator of the expression* (δ^{ω}), p_1 - the *indexical variable of the operator* p_0 or the *variable which accompanies the operator* p_0, p_2 - the *scope of the operator* p_0.

The expression (δ^{ω}) is read: the operator expression token consisting of the main operator p_0, its indexical variable p_1 and its scope p_2. Instead of the term "operator expression token" we shall often use the shorter form "operator expression".

The adopted terminological conventions will be used later in the verbal formulation of further theses of **TSCω-L**.

The following conclusion is a specified version of

Conclusion II.3; it follows from the latter and Definition III.1*.

CONCLUSION III.3

$$p = \rho^{\omega}(p_0,p_1,p_2) \wedge q = \rho^{\omega}(q_0,q_1,q_2) \Rightarrow$$

$$\Rightarrow (p \approx q \Leftrightarrow p_0 \approx q_0 \wedge p_1 \approx q_1 \wedge p_2 \approx q_2).$$

Two operator expressions are equiform if and only if their main operators, the indexical variables of those operators and the scopes of those operators are equiform.

We have the following theorem which is analoguous to Axiom II.10:

THEOREM III.1

$$p = \rho^{\omega}(p_0,p_1,p_2) \Rightarrow (q \approx p \Leftrightarrow$$

$$\Leftrightarrow \bigvee_{q_0,q_1,q_2} (q = \rho^{\omega}(q_0,q_1,q_2) \wedge \bigwedge_{0 \le k \le 2} q_k \approx p_k)).$$

If p is an operator expression then an expression token q is equiform with it if and only if q is also an operator expression and the expressions of its expansion - the main operator, the indexical variable of that operator and the scope of that operator - are equiform with the main operator of p, the indexical variable of that operator and scope of that operator, respectively.

The proof of this theorem is obtained from Definition III.1*, Axioms II.10, III.2 and III.3, Convention II.3a and Conclusion II.4c.

The concept of the *relation (fv) of being a free variable* in a given expression token is a secondary (derived) concept in TSCω-L. That relation is defined by the expression $p(fv)q$, which is read: p is a free variable in an expression token q.

DEFINITION III.1

$$p(fv)q \Leftrightarrow p \in Vr \wedge p \in C_q \wedge$$

$$\wedge \neg \bigvee_{r,r_0,r_1,r_2} (r = \rho^{\omega}(r_0,r_1,r_2) \wedge r \in C_q \wedge p \in C_{r_2} \wedge p \approx r_1).$$

p is a free variable in an expression token q if and only if p is a variable and at the same time a constituent of q and there is no operator expression r such that it is a constituent of q and the variable p is at the same time a constituent of the scope of the main operator of r and a variable equiform with the indexical variable of that operator.

Definition III.1 and Conclusion II.16a immediately yield

CONCLUSION III.4

$$p(fv)q \wedge s \approx q \Rightarrow p(fv)s.$$

If p is a free variable in an expression token q, then it is a free variable in an expression equiform with q.

The concept of a *variable bound by a given operator* is associated with that of a free variable in a given expression. A variable which in a given operator expression is bound by its main operator is a free variable within the scope of that operator and equiform with its indexical variable. Thus in an operator expression p such that $p = \rho^{\omega}(p_0,p_1,p_2)$, p_3 is a variable bound by the main operator p_0 if and only if $p_3(fv)p_2$ and $p_3 \approx p_1$.[24] The operator p_0 is then said to *bind the variable* p_3.

The operator in a given operator expression may happen not to bind any free variable in its scope. We then say that it is *voidly binding* in that operator expression. We also say that an operator is voidly binding in a given operator expression if there is no free variable in its scope. Operator expressions which include voidly binding operators are not included in the well-formed expressions of the ω-language $\mathcal{L}$.

Sec. III.4. Well-formed expressions

The definition of the *set* S *of all well-formed expression tokens* of the ω-language $\mathcal{L}$ will be based on the inductive definition of the set ${}^{n}S$ *of all well-formed expression tokens of order* n ($n \geq 0$). In the symbolic recording of this definition we shall refer to Convention II.3b,c (see Convention III.2). In the verbal formulations the term "well-formed expression token" will be replaced by the shorter term "well-formed expression"

DEFINITION III.2

a. $${}^{0}S = E_s,$$

b. $$p \in {}^{k+1}S \Leftrightarrow p \in {}^{k}S \vee$$

$$\vee \bigvee_{n\geq 1} \bigvee_{p_0, p_1, \ldots, p_n \in {}^{k}S} (p = \rho(p_0, p_1, \ldots, p_n) \wedge p_0 \notin O \wedge$$

$$\wedge\, c^{n+1}(\alpha, \alpha_1, \ldots, \alpha_n, \alpha_0)) \vee \bigvee_{p_0, p_1, p_2, p_3} (p = \rho^{\omega}(p_0, p_1, p_2) \wedge$$

$$\wedge\, p_2 \in {}^{k}S \wedge p_3 (fv) p_2 \wedge p_3 \approx p_1 \wedge c^{3}(\alpha, \alpha_1, \alpha_2, \alpha_0)),$$

c. $$S = \bigcup_{n=0}^{\infty} {}^{n}S.$$

In accordance with Definition III.2a a well-formed expression of order 0 is a simple expression of the ω-language $\mathscr{L}$.

In accordance with Definition III.2b a well-formed expression of order k+1 is an expression which satisfied one of the following three conditions:

1° p is a well-formed expression of order k, and hence an element of the set ${}^{k}S$,

2° p is a compound expression of the ω-language $\mathscr{L}$ such that all elements of its expansion $p_0, p_1, \ldots, p_n$ are well-formed expressions of order k, which is to say that they are elements of the set ${}^{k}S$; further, the main functor p_0 of p is not an operator; finally p satisfies the condition

$$\mathrm{m}_p \qquad \mathrm{c}^{n+1}(\alpha, \alpha_1, \ldots, \alpha_n, \alpha_0) \quad (n \geq 1),$$

in accordance with which the categorial index α_0 of the main functor of p is a concatenation of the index α of p and the indices $\alpha_1, \ldots, \alpha_n$ of all successive arguments of the main functor,

3° p is an operator expression, the main operator p_0 of which binds a certain variable p_3 which is free in its scope p_2 which is a well-formed expression of order k; the index α_0 of that operator is a concatenation of the index α of p, the index α_1 of the indexical variable p_1 of that operator and the index α_2 of its scope p_2; p thus satisfies the condition

$\mathfrak{m}_\rho$ $c^3(\alpha,\alpha_1,\alpha_2,\alpha_0)$.

In accordance with Definition III.2c the set S is the sum of all sets of well-formed expressions of a finite order (greater than or equal to zero).

We now adopt two terminological conventions. A well-formed expression ρ for which there is a natural number k such that ρ satisfies the condition 2° will be termed a *well-formed compound non-operator expression*. A well-formed expression ρ for which there is a natural number k such that ρ satisfies the condition 3° will be termed a ***well-formed operator expression***. We shall use the symbols $S^{n\omega}$ and S^{ω} to denote, the set of all well-formed compound non-operator expressions and the set of all well-formed operator expressions, respectively. The definitions of these concepts are recorded as follows:

Definition III.3

$$\rho \in S^{n\omega} \Leftrightarrow \bigvee_{k} \bigvee_{n\geq 1} \bigvee_{\rho_0,\rho_1,\ldots,\rho_n \in {}^kS} (\rho = \rho(\rho_0,\rho_1,\ldots,\rho_n) \wedge$$
$$\wedge\ \rho_0 \not\approx O \wedge c^{n+1}(\alpha,\alpha_1,\ldots,\alpha_n,\alpha_0)).$$

Definition III.4

$$\rho \in S^{\omega} \Leftrightarrow \bigvee_{k} \bigvee_{\rho_0,\rho_1,\rho_2,\rho_3} (\rho = \rho^{\omega}(\rho_0,\rho_1,\rho_2) \wedge \rho_2 \in {}^kS \wedge$$
$$\wedge\ \rho_3(fv)\rho_2 \wedge \rho_3 \approx \rho_1 \wedge c^3(\alpha,\alpha_1,\alpha_2,\alpha_0)).$$

Note that the concept of the set $S^{n\omega}$ of all well-formed compound non-operator expressions is in **TSCω-L** an analogon of the concept of the set $S\setminus{}^0S$ of all well-formed compound expressions, introduced in **TSCL** (see Sec.II.5). It must further be noted that a non-operator expression need not be an operator-free one: in the expansion of a well-formed compound non-operator expression there may occur, as an argument of the main functor of that expression, an operator expression or an expression which has an operator expressions as its constituent.

In formulating this remark, and earlier the Condition 2°, we have availed ourselves of the fact that every well-formed compound non-operator expression is a compound expression. Definition III.3,

LEMMA III.1

$$^{n}S \subseteq E \subseteq D(\iota),$$

Definition II.3b and Axiom II.9 yield

CONCLUSION III.5a

$$S^{n\omega} \subseteq E_{C}.$$

Lemma III.1 is an analogon of Lemma II.2. Its proof is given in the Annex.

Definitions III.4, III.1*, Conclusions III.1a,b and II.5b, Definition II.3b and Axiom II.9 yield also

CONCLUSION III.5b

$$S^{\omega} \subseteq E_{C}.$$

Every element of the set S^{ω} *is a compound expression token.*

Note that the main functor of each expression from the set S^{ω}, unlike the main functor of any expression from the set $S^{n\omega}$, is an operator. The sets $S^{n\omega}$ and S^{ω} are thus disjoint because ρ is a one-to-one function (Axiom II.9). Each of these sets is also disjoint with the set E_S, which follows from Conclusions III.5a,b and from the fact that, by Definition II.3a and Conclusion II.5a. $E_C \cap E_S = \emptyset$.

We adopt an axiom analoguous to Axiom II.11, namely

AXIOM III.4

$$\iota(S^{n\omega}) \cap I_o \neq \emptyset.$$

In the ω-language $\mathcal{L}$ there is a well-formed compound non-operator expression which has a basic categorial index.

We also adopt

AXIOM III.5

$$\iota(S^{\omega}) \cap I_o \neq \emptyset.$$

In the ω-language $\mathcal{L}$ there is a well-formed operator expression which has a basic categorial index.

The sets $S^{n\omega}$ and S^{ω} are non-empty in accordance with Axioms III.4 and III.5. The same applies to the set E_S because otherwise, by Definitions III.2a,b, each of the sets ${}^{k}S$ would be empty, too, and thus, in view of Definitions III.3 and III.4, the sets $S^{n\omega}$ and S^{ω} would be empty.

Moreover, it follows from Definitions III.2c,a,b, III.3 and III.4 that the sum of E_S, $S^{n\omega}$ and S^{ω} equals S. This and the earlier analyses yield

THEOREM III.2

$$S = {}^{0}S \cup S^{n\omega} \cup S^{\omega} \wedge {}^{0}S \neq \emptyset \wedge S^{n\omega} \neq \emptyset \wedge S^{\omega} \neq \emptyset \wedge$$

$$\wedge\ {}^{0}S \cap S^{n\omega} = \emptyset \wedge {}^{0}S \cap S^{\omega} = \emptyset \wedge S^{n\omega} \cap S^{\omega} = \emptyset.$$

The set of all well-formed expressions of the ω-language $\mathcal{L}$ *is the sum of three non-empty and pairwise disjoint sets: the set* ${}^{0}S$ *of all simple expressions of the ω-language* $\mathcal{L}$, *the set* $S^{n\omega}$ *of all well-formed compound non-operator expressions of that language and the set* S^{ω} *of all well-formed operator expressions of that language.*

This theorem, Definition III.2c and Lemma III.1 yield immediately

CONCLUSION III.6

$$\emptyset \neq S \subseteq E \wedge S \subseteq D(\iota).$$

The set of all well-formed expressions of the ω-language $\mathcal{L}$ *is a non-empty set of expression tokens of that language and the set of word tokens which have an index.*

Since S^{ω} is non-empty it follows from Definition III.4 and Metadefinition III.1* that O and Vr are non-empty, too. By Axiom III.1 and Definitions III.2a,c these sets are subsets of S. Thus we arrive at

CONCLUSION III.7

a. $\emptyset \neq O \subseteq S,$

b. $\emptyset \neq Vr \subseteq S.$

In accordance with Conclusions III.7a,b the set of all operators and the set of all variables are non-empty subsets of the set of all well-formed expressions of $\mathscr{L}$.

Sec. III.5. Fundamental theorems

In this section we shall formulate theorems pertaining to the well-formed expressions of ω-languages. They are analoguous to the theorems of **TSCL** pertaining to well-formed expressions of simple categorial languages.

Theorem II.3 has its analogon in

THEOREM III.3

$$p \in S \wedge q \approx p \Rightarrow q \in S.$$

An expression of the ω-language $\mathscr{L}$ which is equiform with its well-formed expression is also a well-formed expression.

This theorem follows from Definition III.2c and the lemma which is analoguous to Lemma II.3:

LEMMA III.2

$$p \in {}^{k}S \wedge q \approx p \Rightarrow q \in {}^{k}S.$$

An expression equiform with a well-formed expression of order k is also a well-formed expression of order k.

The inductive proof of this lemma is given in the Annex. This lemma is used in the proofs of the following two theorems:

THEOREM III.3a

$$p \in S^{n\omega} \wedge q \approx p \Rightarrow q \in S^{n\omega}.$$

An expression equiform with a well-formed compound non-operator expression is also a well-formed compound non-operator expression.

THEOREM III.3b

$$p \in S^{\omega} \wedge q \approx p \Rightarrow q \in S^{\omega}.$$

An expression equiform with a well-formed operator expression is also a well-formed operator expression.

The proofs of Theorems III.3a,b, based on Lemma III.2, are quite similar to the proof of that lemma in the cases 2° and 3° on the basis of the inductive assumption (see Annex). They are therefore omitted.

In view of Convention II.3c, Conclusions III.6 and II.2b yield

CONCLUSION III.8

$$p,q \in S \wedge p \approx q \Rightarrow a \approx b.$$

The indices of equiform well-formed expressions are equiform.

We shall now formulate definitions which are analoga of Definitions II.6 and II.7. They are definitions of two concepts: the *set B of all basic expressions* of the ω-language $\mathcal{L}$ and the *set F of all functors* of that language.

DEFINITION III.5

$$B = \{ p \in S \mid a \in I_{o} \}.$$

The set of all basic expressions is the set of all those well-formed expressions of the ω-language $\mathcal{L}$ which have a basic index.

Definition III.6

$$F = \{ p \in S \mid \alpha \in I \setminus I_0 \}.$$

The set of all functors is the set of all well-formed expressions of the ω-language ℒ which have an index that is not a basic index.

Definition III.5, Theorem III.3, Conclusion III.8 and Axiom II.2 yield

Theorem III.3c

$$p \in B \wedge q \approx p \Rightarrow q \in B.$$

An expression equiform with a basic expression is also a basic expression.

Definition III.6, Theorem III.3, Conclusion III.8, Theorem I.4^{d}a and Axiom I.7 yield

Theorem III.3d

$$p \in F \wedge q \approx p \Rightarrow q \in F.$$

An expression equiform with a functor is also a functor.

The definitions of B and F make it possible to formulate a theorem which is an analogon of Theorem II.4 in TSCL. Before we record it we shall first draw several conclusions from these definitions.

Axiom III.4, Theorem III.2 and Definition III.5 yield immediately the following conclusion which is analoguous to Conclusion II.10:

CONCLUSION III.9

$$S^{n\omega} \cap B \neq \emptyset.$$

There is a well-formed compound non-operator expression of the ω-language $\mathcal{L}$ which is a basic expression.

Axiom III.5, Theorem III.2 and Definition III.5 yield

CONCLUSION III.10

$$S^{\omega} \cap B \neq \emptyset.$$

There is a well-formed operator expression which is a basic expression.

Since, by Conclusion III.10, there is a well-formed operator expression, and, by Definitions III.4 and III.1*, each such expression consists of an operator whose index is a three-element concatenation of indices, by Theorem I.13^{d}b (see Conventions III.2 and II.2 and Conclusion III.2*) there is an operator whose index belongs to the set $I \setminus I_o$. Thus, by Conclusion III.7a and Definition III.6, we have

CONCLUSION III.11

$$O \cap F \neq \emptyset.$$

At least one operator of the ω-language $\mathcal{L}$ is a functor.

This conclusion is at variance with the often adopted assumption that an operator cannot be a functor. It is, however, a cosequence of the here adopted definition of F. The disjointness of O and F could, of course, be guaranteed by the adoption of a different definition of F, for instance

$$F = \{ \rho \in S \setminus O \mid \alpha \in I \setminus I_o \}.$$

Since $S^{n\omega}$ is a non-empty subset of S (Theorem III.2), by Definitions III.3 and III.2c, Theorem I.13^{d}b and Definition III.6 we arrive at

CONCLUSION III.12

$$F \setminus O \neq \emptyset.$$

There is a functor which is not an operator.

We shall record one more conclusion pertaining to functors.

CONCLUSION III.13

$${}^{0}S \cap F \neq \emptyset.$$

There is a simple expression of the ω-language $\mathcal{L}$ which is a functor.

Conclusion III.13 is an analogon of Conclusion II.11. It can be proved by reference to Conclusion III.11, Axiom III.1 and Definition III.2a.

Conclusions III.11–III.13 do not, of course, preclude the possibility of including operators in the set of functors. It follows from each of them that F is non-empty. On the other hand, the non-emptiness of R follows immediately from Conclusion III.9. This, Definitions III.5, III.6, Conclusion III.6 and Axiom II.7 yield immediately the afore-mentioned analogon of Theorem II.4:

Theorem III.4

$$S = B \cup F \wedge B \neq \emptyset \wedge F \neq \emptyset \wedge B \cap F = \emptyset.$$

The set of all well-formed expressions of the ω-language $\mathcal{L}$ is the sum of two non-empty and disjoint sets, that of all basic expressions of that language and that of all its functors.

Note that if we adopted a definition of F other than Definition III.6, inserted earlier, and an additional axiom: $\iota(O) \subseteq I \setminus I_O$, we could replace Theorem III.4 by the following one:

The set of all well-formed expressions of the ω-language $\mathcal{L}$ is the sum of three non-empty and pairwise disjoint sets: that of all basic expressions of that language, that of all its functors, and that of all its operators.

The ω-language $\mathcal{L}$ under consideration in this chapter is a *syntactically categorial language*. The set S of all its well-formed expressions can be split into syntactic categories so that every well-formed expression belongs to exactly one syntactic category. Before we formulate the appropriate theorem which is an analogon of Theorem II.6a in **TSCL** we note that **TETk** includes the theorem which states that the relation $\underset{c}{=}$ of categorial agreement is an equivalence relation in E (see Conclusions II.13a-c) and also the theorem which enables us to identify syntactic categories with the appropriate classes of abstraction of that relation (see Conclusion II.14). Since, by Conclusion III.6, the set E is non-empty (earlier that fact was proved by reference to the means available in **TSCL**: Conclusion

II.9d follows from Conclusions II.6a,b and II.9a,b), Theorem II.6 is also a theorem in **TSCω-L**. The family Ct of all syntactic categories is thus a logical partition of E determined by the relation $\underset{c}{\equiv}$. That relation, of course, also determines a logical partition of every non-empty subset of E. By Conclusion III.6 and Theorem III.4, the sets S, B, and F, defined in **TSCω-L**, are non-empty subsets of E. The following subfamilies of the family Ct are, respectively, logical partitions of those sets: the family $Ct(S)$ of all syntactic categories of the well-formed expressions of the ω-language $\mathcal{L}$, the family $Ct(B)$ of all basic categories of expressions of that language, and the family $Ct(F)$ of all functor caregories of expressions of that language. The definitions of those concepts are analoguous to Definitions II.11a-c:

DEFINITION III.7

a. $Ct(S) = \{ Ct_{\xi} \mid \xi \in \iota(S) \}$,

b. $Ct(B) = \{ Ct_{\xi} \mid \xi \in \iota(B) \}$,

c. $Ct(F) = \{ Ct_{\xi} \mid \xi \in \iota(F) \}$.

The comments made above allow us to formulate

THEOREM III.5. *The relation $\underset{c}{\equiv}$ determines a logical classification (logical partition) of the set S of all well-formed expressions of the ω-language $\mathcal{L}$ into syntactic categories which are non-empty, pairwise disjoint and such that their sum equals S. Further, S can be presented as the sum consisting of the sum of all basic categories equal to the set B and the sum of all functor categories equal to the set F.*

Thus, the following conditions are satisfied:

(i) $\bigwedge_{Ct_\xi \in Ct(S)} Ct_\xi \neq \emptyset,$

(ii) $\bigwedge_{Ct_\xi, Ct_\sigma \in Ct(S)} (Ct_\xi \neq Ct_\sigma \Rightarrow Ct_\xi \cap Ct_\sigma = \emptyset),$

(iii) $\bigcup_{Ct_\xi \in Ct(S)} Ct_\xi = S,$

(iv) $S = \bigcup_{Ct_\xi \in Ct(B)} Ct_\xi \cup \bigcup_{Ct_\xi \in Ct(F)} Ct_\xi \wedge$

$$\wedge \bigcup_{Ct_\xi \in Ct(B)} Ct_\xi = B \wedge \bigcup_{Ct_\xi \in Ct(F)} Ct_\xi = F.$$

This section will be concluded by the formulation of two *fundamental theorems of the theory of syntactic categories*. Those theorems, for which we adopt the abbreviations I **fttsc** and II **fttsc**, are based on the expression (see Sec.II.1 and Sec.II.7) $r(p/q)s$, which is read: the expression r is obtained from the expression s by the replacement of the constituent q of the latter by the expression p. In defining it we avail ourselves, as in the case of **TSCL**, of the inductive definition of the expression $r(p/q)^n s$, which denotes here the relation of replacement of a constituent of n-th order of a given expression of the ω-language $\mathcal{L}$.

DEFINITION III.8

a. $r(p/q)^0 s \Leftrightarrow s,r \in S \wedge q \approx s \wedge p \approx r,$

b. $r(p/q)^1 s \Leftrightarrow (s,r \in S^{n\omega} \wedge$

$$\wedge \bigvee_{n\geq 1} \bigvee_{\substack{s_0,s_1,\ldots,s_n \in S \\ r_0,r_1,\ldots,r_n \in S}} (s=\rho(s_0,s_1,\ldots,s_n) \wedge r=\rho(r_0,r_1,\ldots,r_n) \wedge$$

$$\wedge \bigvee_{0\leq j\leq n} (q \approx s_j \wedge p \approx r_j \wedge \bigwedge_{\substack{k\neq j \\ 0\leq k\leq n}} s_k = r_k))) \vee (s,r \in S^{\omega} \wedge$$

$$\wedge \bigvee_{\substack{s_0,s_1,s_2 \\ r_0,r_1,r_2}} (s = \rho^{\omega}(s_0,s_1,s_2) \wedge r = \rho^{\omega}(r_0,r_1,r_2) \wedge$$

$$\wedge ((q \approx s_0 \wedge p \approx r_0 \wedge s_1 \approx r_1 \wedge s_2 \approx r_2) \vee (q \approx s_1 \wedge p \approx r_1 \wedge$$

$$\wedge \bigwedge_{0\leq k\leq 2} s_k \approx r_k) \vee (q \approx s_2 \wedge p \approx r_2 \wedge s_0 \approx r_0 \wedge s_1 \approx r_1)))),$$

c. $k > 0 \Rightarrow (r(p/q)^{k+1} s \Leftrightarrow \bigvee_{t,v} (r(v/t)^k s \wedge v(p/q)^1 t)),$

d. $r(p/q)s \Leftrightarrow \bigvee_n r(p/q)^n s.$

Definitions III.8a,c,d have the same forms and verbal formulations as Definitions II.14a,c,d, respectively. In accordance with Definition III.8b the expression r is obtained from the expression s by the replacement of its constituent q of the first order by the expression p if and only if either

1° s and r are well-formed compound non-operator expressions with expansions consisting of the same number of well-formed expressions of the ω-language $\mathcal{L}$ and such that all components with the same numbers other that a certain j are equiform, and the j-th expression of the expansion of s is equiform with q

and the j-th expression of the expansion of r is equiform with p, or

2° s and r are wel-formed operator expressions such that one of the three following conditions is satisfied:

(a) q is equiform with the main operator of s, p is equiform with the main operator of r, and the indexical variables of those operators and their scopes are equiform,

(b) q is equiform with the indexical variable of the main operator of s, p is equiform with the indexical variable of the main operator of r, and the main operators of s an r and the indexical variables and the scopes of those operators are equiform,

(c) q is equiform with the scope of the main operator of s, p is equiform with the scope of the main operator of r, and the main operators of s and r and their indexical variables are equiform.

Note that in the case 2°(b) if the replaceable expressions are equiform with the corresponding indexical variables, then those variables are equiform. In such a case s and r are, by Conclusion III.3, equiform, too.

Note also that on the strength of Definitions III.8a-d we can arrive at conclusions whose symbolic recording does not differ from that of Conclusions II.20a,b, II.21-II.23.

We now pass to the two **fttsc** which are essential for **TSCω-L**. Their graphic form does not differ from the two **fttsc** valid in **TSCL** (Theorems II.7a and II.7b).

Theorem III.6a (I **fttsc**)

$$r(p/q)s \wedge r \underset{c}{=} s \Rightarrow p \underset{c}{=} q.$$

If an expression r of the ω-language ℒ is obtained from an expression s of that language, which belongs to the same syntactic category, by the replacement of its constituent q by an expression p, then the replaceable expressions p and q also belong to the same syntactic category.

The proof of this theorem, modelled after the proof of I **fttsc** in Sec.II.7, is given in the Annex.

Theorem III.6b

$$r(p/q)s \wedge p \underset{c}{=} q \Rightarrow r \underset{c}{=} s.$$

If an expression r of the ω-language ℒ is obtained from an expression s of that language by the replacement of its constituent q by an expression p which belongs to the same syntactic category as q, then r and s also belong to the same syntactic category.

The proof of Theorem III.6b can be easily reconstructed if we follow the proof of II **fttsc** given in Sec.II.7, and also the proof of Theorem III.6a. It is therefore omitted here.

The two **fttsc** taken jointly enable us to prove

Theorem III.6

$$r(p/q)s \Rightarrow (p \underset{c}{=} q \Leftrightarrow r \underset{c}{=} s).$$

Two expressions of the ω-language ℒ belong to the same syntactic category if and only if by replacing one by the other in a well-formed expression of the ω-language ℒ we obtain a well-formed expression of that language which belongs to the same syntactic category as the latter.

Sec. III.6. The algorithm of checking the syntactic correctness of expressions

The algorithm to be described here pertains to compound expressions of the ω-language $\mathscr{L}$. It is similar to the algorithm given in Sec.II.8 and pertaining to the compound expressions of any simple categorial language.

The preliminary assumptions concerning the symbolism of the ω-language $\mathscr{L}$, analysed in this chapter, are an expansion of the assumptions formulated in Sec.II.8. We accordingly assume that every label consists of a finite number of symbols, that we are able, referring to empirical criteria, to decide about any two labels whether or not, they are equiform and about any three labels, whether one of them is a concatenation of the remaining two. We also assume that we can decide about any label whether it is in the vocabulary, whether it is a word that has an index, whether it is an operator, or a variable, or a compound expression. If it is a compound expression, we are in a position to establish which parts of it correspond to the words of its expansion, which part of it corresponds to the main functor, and which parts of it correspond to the successive arguments of that functor. We are also able to assign a categorial index to every expression of the ω-language $\mathscr{L}$. We also assume that about any sequence of indices

(ξ) $\qquad \xi_0, \xi_1, \xi_2, \ldots, \xi_n \quad (n \geq 1)$

we can decide whether it satisfies the condition

$\mathfrak{m}_p$ $\qquad c^{n+1}(\alpha, \xi_1, \xi_2, \ldots, \xi_n, \xi_0)$,

where α is the index of the expression p.

We also retain the term "intermediate expansion of a compound expression". Let us recall that the sequence (ξ) is an intermediate expansion of the expression

(δ) $\qquad \rho(p_0, p_1, \ldots, p_n)$

such that $p_0, p_1, \ldots, p_n$ are expressions if and only if for any i such that $0 \leq i \leq n$, $\xi_i \approx \alpha_i$ and α_i is the index of the i-th element of the expansion of (δ).

Thus if a compound expression of the ω-language $\mathscr{L}$ is an operator expression of the form

(δ^{ω}) $\qquad \rho^{\omega}(p_0, p_1, p_2)$,

then, in accordance with the convention adopted and Definition III.1*, the intermediate expansion of (δ^{ω}) is any sequence of indices

(ξ^{ω}) $\qquad \xi_0, \xi_1, \xi_2$

such that for any i, $0 \leq i \leq 2$, $\xi_i \approx \alpha_i$, where $\alpha_i = \iota(p_i)$.

The algorithm of checking the syntactic correctness of a compound expression of the ω-language $\mathscr{L}$ begins with an auxiliary algorithm, namely *that of constructing tree* I of that expression (see Sec.II.8). In carrying out the successive operations of that algorithm we find whether tree I of a given compound expression is, or is not, a tree of a *thoroughly well-arranged expression*. If the answer is in the affirmative, then we order the elements of that tree in the manner described in Sec.II.8, and next we pass to the successive algorithm, namely *that of checking the existence of voidly binding operators* in the expression under consideration.

In accordance with that algorithm we first of all check, in the order fixed for the elements of tree I of the expression in question, whether at least one of its final nodes is an operator. If this is not the case, then the expression in question, which is thoroughly well-arranged, does not, of course, include voidly binding operators, and since it can be included in that fragment of the ω-language $\mathcal{L}$ which satisfies the conditions imposed upon simple syntactically categorial languages we submit it to further operations described in Sec.II.8. If, on the contrary, at least one of the final nodes of tree I of the expression in question is an operator, then we transform tree I of that expression into *tree* I' by proceeding in accordance with the following directives:

1° In the order fixed for the elements of tree I we underline all variables which are the final nodes of tree I, of the expression in question; those variables are their own free variables.

2° Once we know all the free variables which occur, and are underlined, in the successive expression - which have no voidly binding operators - in the *k*-th line of tree I being subject to transformation, we find, while observing the order from the left to the right, whether in any compound expression p in the k-1-th line situated above in that tree there is a voidly binding operator, and we establish what possible further operations should be made in transforming tree I.

(a) If the main functor of p is not an operator, then p does not include a voidly binding operator. Then in every part of that expression which corresponds to the component p_i (i>0) of the expansion of p we underline every such variable which is equiform with the free variable that occupies the same syntactic position in that expression of the k-th line, which expression is equiform with p_i. The variables underlined in that manner are all free variables in p.

(b) If the main functor of p is an operator and p has its expansion in the sequence p_0,p_1,p_2, then in the k-th line there occur expressions q_0,q_1,q_2, equiform, respectively, with p_0,p_1,p_2 and such that either

(b_1) q_1 is not equiform with any free variable in q_2 or there is no free variable in q_2, or

(b_2) q_1 is equiform with a variable q_3 which is free in q_2.

In the case(b_1) the operator p_0 in p is voidly binding and tree I' of the expression under consideration has been constructed.

In the case (b_2) the operator p_0 in p binds a variable which is equiform with the variable q_3 and is within its scope p_2. Then p has no voidly binding operators. In that case in that part of p which corresponds to p_2 we underline every such variable which is equiform with a free variable in q_2, other than q_3, which has the same syntactic position. The variables underlined in ρ are all free variables in ρ.

3° The construction of tree I' is terminated when, in carrying operations 1° and 2°, we encounter a voidly binding operator or we reach the first line of tree I, which is the expression under consideration, and find that it has no voidly binding operator.

If in tree I' of the expression under consideration there is an element which includes a voidly binding operator the algorithm of checking the well-formedness of that expression has been completed: the expression in question is not well-formed. If tree I' proves to be a tree of an expression which has no voidly binding operator, then we pass to the successive operations of checking the syntactic correctness of that expression.

For that purpose, in accordance with the fixed order of elements of tree I we successively replace its elements by their categorial indices. We obtain in this way the *tree of categorial indices* of a given compound expression (see Sec.II.8). That tree will be termed *tree* II of the expression being checked.

As in Sec.II.8, we proceed to order the intermediate expansions of the compound expressions of tree I, formed of parts of tree II. Out of two intermediate expansions that one is the *earlier* which is an intermediate expansion of an earlier element of tree I.

The description of the last part of the algorithm of

checking the well-formedness of a thoroughly well-arranged compound expression which does not include voidly binding operators does not differ from the description given in Sec.II.8. In accordance with the order adopted for intermediate expansions we check whether, for any compound expression ρ which is an element of tree I, those parts of tree II which form the intermediate expansion of the expression ρ satisfy the condition m_ρ. When we find the first intermediate expansion which does not satisfy the condition m_ρ the algorithm of checking the syntactic correctness has been completed: the expression being checked is not well-formed. If the intermediate expansion of any compound expression ρ of tree I satisfies the desired condition, then the algorithm of checking the well-formedness of the expression in question has been completed, too: the expression being checked is well-formed.

In the examples given below the concatenations of the indices which correspond to the indices of the main functors (including main operators) of compound expressions which are elements of tree I are written in the linear quasi-fractional form, as was the case in Sec.II.8. The index which occurs before the slant / indicates the index of the expression which a given functor forms; the indices which occur after that symbol indicate the indices of the successive arguments of that functor. To a sentential expression we assign an index which is equiform with *s*. To the index of a quantifier which binds a name variable that has an index *n* we accordingly assign a label which is equiform with *s/ns*, and to the index of a quantifier which binds a sentential variable we assign a label which is equiform with *s/ss*. About the former label we say that it is the index of a sentence-forming functor of two arguments, the first of which is a name variable, and the second is a sentential expression which contains a name variable that is free in it. In other words, it is the index of a functor which together with a name variable and a sentential expression which contains that name variable that is free in it forms a sentential expression. Likewise, we say about the second label that it is the index of a functor which together with a sentential variable and a sentential expression that contains a sentential variable which is free in it forms a sentential expression.

The above description of the indices of operators makes it possible to bring out the specific character of those functors which are operators and to distinguish them from "ordinary" functors.

We shall now apply this algorithm of checking well-formedness to three compound arithmetical expressions. In the recording of those expressions we use brackets in the way their are ordinarily used.

E x a m p l e III.1. We shall first check the syntactic correctness of the expression

(π1) $$\bigwedge_{y} \bigvee_{y} x > y.$$

In order to find whether the expression (π1) is thoroughly well-arranged we construct tree I of that expression

(Fig.III.1). All terminal nodes of that tree are simple expressions, and hence (π1) is thoroughly well-arranged. We accordingly transform tree I into tree I' and see if (π1) includes voidly binding operators (see Fig.III.1).

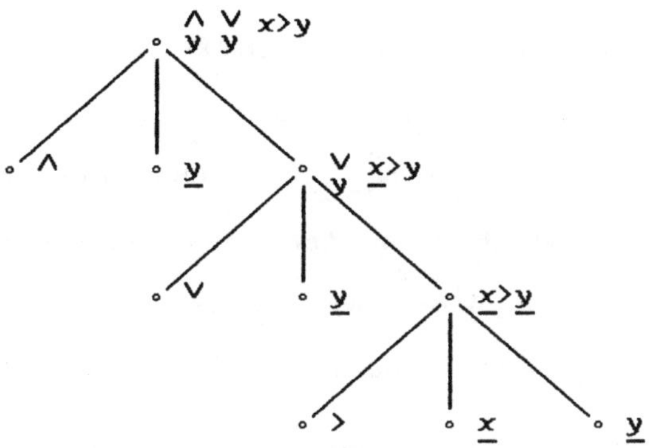

Fig.III.1. Tree I' of the expression (π1)

Note that the existential quantifier which occurs in the expression in the second line binds a variable which is equiform with the variable y. On the contrary, the universal quantifier in the first line of the tree is voidly binding because the variable which occurs in the second line in the second place is not equiform with any free variable in the expression which occurs in that line in the third place. Hence the expression being checked is not well-formed.

E x a m p l e III.2. Let us check now the well-formedness of the expression

(π2) $$\bigwedge_x \bigvee_y x + (y \cdot z)$$

We construct tree I and then tree I' of (π2) (Fig.III.2). It can be seen that the expression being checked is thoroughly well-arranged and does not include voidly binding operators. We must accordingly construct tree II of (π2) and see if for any compound expression ρ of tree I those parts of tree II which form the intermediate expansion of ρ satisfy the condition $\mathfrak{m}_{\rho}$. Tree II of (π2) is shown in Fig.III.3.

Those parts of tree II which occur in its second line and form the intermediate expansion of an expression equiform with (π2) satisfy the condition $\mathfrak{m}_{(\pi 2)}$. On the contrary, the indices in the third line which form the intermediate expansion of a compound expression equiform with $\bigvee_{y} x+(y \cdot z)$ do not satisfy the condition $\mathfrak{m}_{\bigvee_{y} x+(y \cdot z)}$: the index of the existential quantifier of that expression is not a concatenation of the index of that expression and, successively, the index of the indexical variable of that quantifier and the index of its scope. Hence the expression being checked is not well-formed.

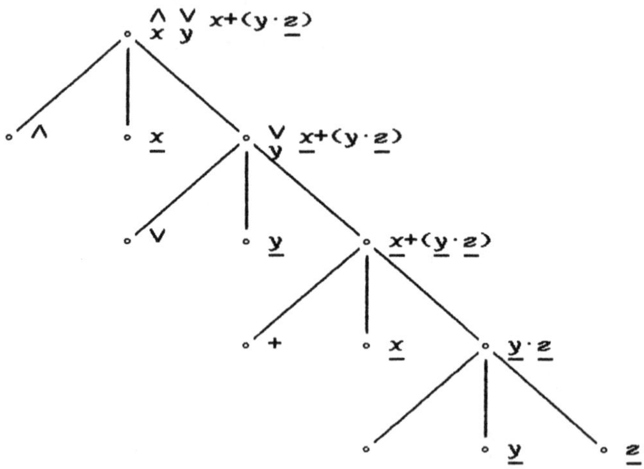

Fig.III.2. Tree I' of the expression (π2)

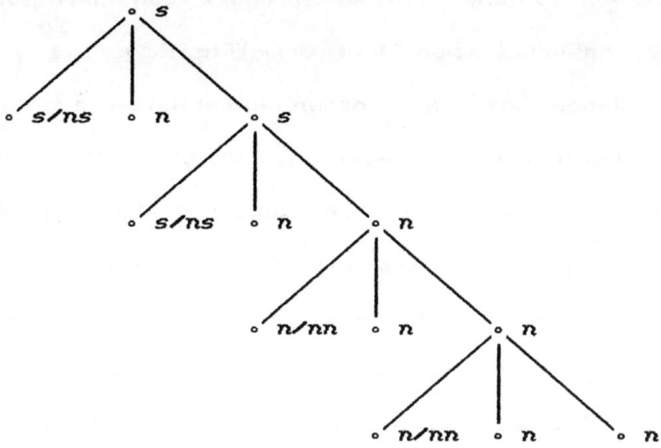

Fig.III.3. Tree II of the expression (π2)

E x a m p l e III.3. We shall check now the syntactic correctness of the expression

(π3) $$\bigwedge_{x} x > 0 \Rightarrow \bigwedge_{y} \bigvee_{z} x + y < z,$$

which has the functor of implication as its main functor. We first construct tree I of that expression and find that (π3) is thoroughly well-arranged. We accordingly transform tree I of that expression into I' (see Fig.III.4).

The operator expression which occurs in the third line and is equiform with $\bigvee_{z}$ x+y<z does not include any voidly binding operator. The two operator expressions which occur in the second line of tree I' and are, respectively, equiform with $\bigwedge_{x}$ x>0 and $\bigwedge_{y}$ $\bigvee_{z}$ x+y<z also do not include voidly binding operators. Hence (π3) is not only thoroughly well-arranged but does not include voidly binding operators, either. We must, therefore, perform the successive operations of checking the

well-formedness of the expression under consideration. For that purpose we construct tree II of (π3) (Fig.III.5).

In accordance with the order established for intermediate expansions we check, successively, whether for any compound expression ρ of tree I the intermediate expansion of ρ determined by parts of tree II satisfies the condition $\mathfrak{m}_{\rho}$. It turns out that each of the seven intermediate expansions of the compound expressions of tree I satisfies that condition. Hence (π3) is a well-formed expression.

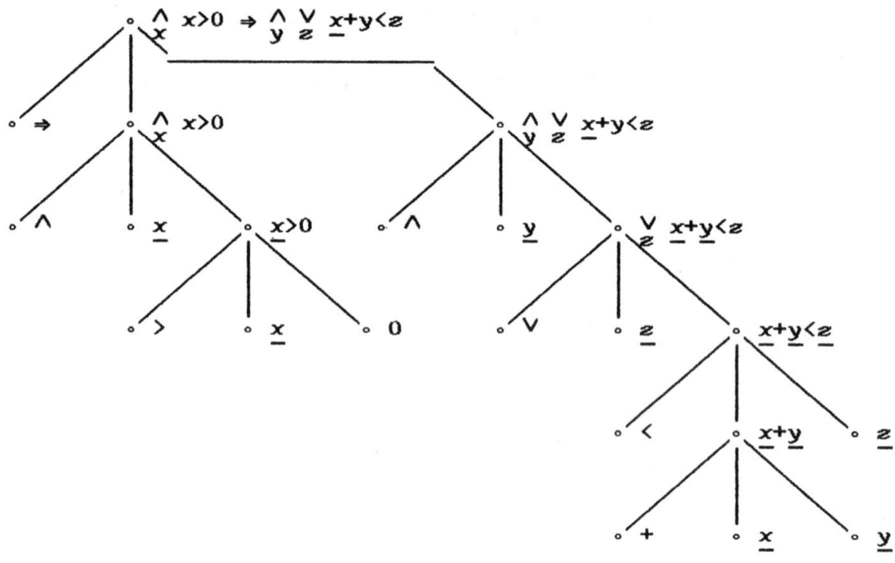

Fig.III.4. Tree I' of the expression (π3)

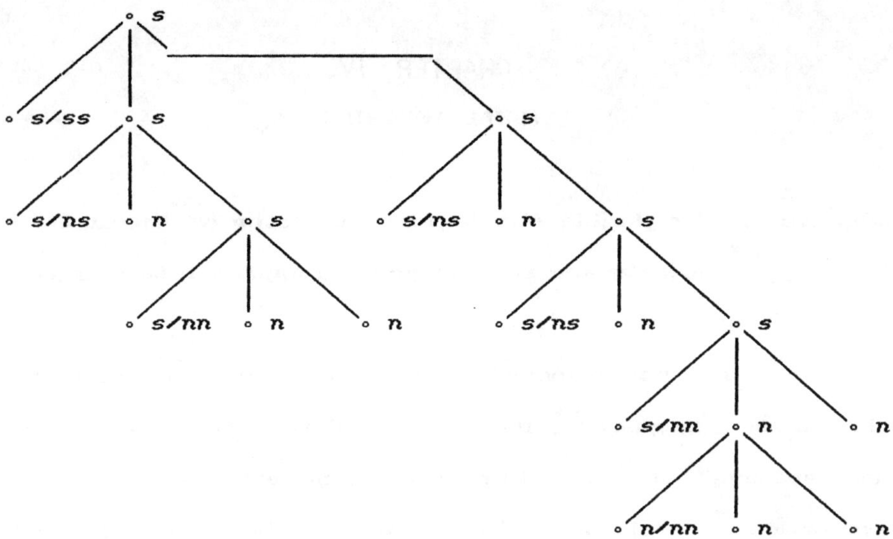

Fig.III.5. Tree II of the expression (π3)

CHAPTER IV

DUAL THEORIES

Sec. IV.1. The double ontological character of linguistic objects and the biaspectual approach to language

There are discrepancies in the semiotic interpretation of the term "language". They are related to the ambiguous understanding of such linguistic components as label, word, expression. Particularly well known is the distinction, originating from C.S.Peirce [1931-35], between labels, words, and expressions as *tokens* and as *types* (also: events and designs, Lat. in concreto and in specie).

Tokens are concrete material objects, *types* are abstract entities which are classes of *tokens*. These two kinds of linguistic objects, belonging to two different ontological categories, are used independently as convenient instruments of semiotic and logical analysis. This is so because in semiotics and logic we refer, on an equal footing, to concrete linguistic objects and to abstract objects. For instance, in the pragmatic analyses concerned with the functional approach to language (see J.Pelc [1967]), which is connected with the use of expressions in definite contexts and situations, we are interested in concrete expressions, i.e., expression tokens. On the other hand, the functioning of such semiotic concepts as meaning, truth, reference, denotation, communication process is based on the concept of expression type[25], i.e., an abstract linguistic product which is the class of all equiform or equiphonic expression tokens. Such an interpretation of expression types seems to be the most natural of all. Such an interpretation of expressions can be seen, more or less clearly, in the contemporary literature concerned with language analysis. The doubt, however, arises whether expression types of graphic languages are always understood as classes of abstraction of expression tokens determined by the relation of equiformity. Many authors do not explain how they interpret labels, words, expressions, even though it follows from the context that they refer to them as types without, however, making use of such concepts as equiformity.

This duality in using the terms "label", "word", "expression" does not impose any definite logical conception of language. Theoretical constructions encountered in the literature of the subject and pertaining to language usually take into account that aspect of language which makes it possible to speak about its expressions as types.

In the preceding three chapters we have been interested in a different point of view. We have been concerned with the formal determination of the properties of label tokens, word tokens, and expression tokens of graphic languages. These syntactic components of languages, being perceivable by sight are concrete objects, that is, tokens. The theories **TLTk**, **TETk**, **TSCL**, **TSCω-L** thus describe language as a product of a different kind, as a language of expression tokens.

The double ontological character of linguistic objects and the double way of making use of them in logical linguistic analysis speak, however, in favour of such theoretical conceptions of language which could explain its double aspects. If one does not want to break away from semiotic practice one should not confine oneself to treating language as the language of expression types only or as the language of expression tokens only. Both aspects of language should be reflected in a single theoretical conception. That goal can be attained. We have precedents in A.A.Markov's [1954] theory of algorithms and also in the theory of concrete and abstract words, inspired by Jerzy Słupecki and presented by G.Bryll and S.Miklos [1977]; the latter, unlike Markov's theory, is axiomatic.

In the present chapter we shall outline several theories whose principal specific feature is that in the description of the object of inquiry they take into account, directly or indirectly, the said biaspectual approach to syntactically categorial languages. These theories will be termed *dual*.

Each of the four dual theories analysed in this chapter, namely **TLTp**, **TETp**, **DTSCL**, and **DTSCω-L**, is definitional extension of the corresponding theory discussed in one of the preceding chapters. Thus, the *theory* **TLTp** *of label types* is a definitional extension of the theory **TLTk** of label tokens; the *theory* **TETp** *of expression types* is a definitional extension of the theory **TETk** of expression tokens[26]; the *dual theory* **DTSCL** *of simple categorial languages* is a definitional extension of the theory **TSCL**; and the *dual theory* **DTSCω-L** *of categorial*

ω-***languages*** is a definitional extension of the theory **TSCω-L**.

The definitions which are added to the theories **TLTk**, **TETk**, **TSCL** and **TSCω-L** to produce the corresponding dual theories make it possible to look at language not as a product earlier characterized by the ordered sixtuple (see Sec.I.1 and Sec.II.1)

$(\mathscr{L})$ $\langle\, Lb;\ V,\ W;\ \approx,\ c;\ S\,\rangle,$

but as a product whose syntactic characteristic is described by the ordered sixtuple

$(\overline{\mathscr{L}})$ $\langle\, \overline{Lb};\ \overline{V},\ \overline{W};\ =,\ \overline{c};\ \overline{S}\,\rangle,$

consisting of the universal *set* $\overline{Lb}$ *of all label types, the vocabulary* $\overline{V}$ *of simple word types, the set* $\overline{W}$ *of all word types, the relation* = *of equality in* $\overline{Lb}$, *the relation* $\overline{c}$ *of concatenation in* $\overline{Lb}$, and the *set* $\overline{S}$ *of all well-formed expression types.*

The sets $\overline{Lb}$, $\overline{V}$, $\overline{W}$, $\overline{S}$ are defined as respective quotient families $Lb/_{\approx}$, $V/_{\approx}$, $W/_{\approx}$, $S/_{\approx}$ that is as the respective quotients of the sets Lb, V, W, S by the relation of equiformity $\approx$. Hence any element of each of those sets is a class of abstraction (equivalence class) of the relation of equiformity $\approx$, determined by a given label token. The relation = is the ordinary relation of extensional equality of sets. It holds between label types, that is, between classes of abstraction of label tokens. The relation of concatenation $\overline{c}$ is a three-argument (ternary) relation which holds between label types.

The concepts $\overline{Lb}$, $\overline{V}$, $\overline{W}$, and $\overline{c}$, which correspond, respectively, to the concepts Lb, V, W, and c in **TLTk**, are defined in **TLTp** (Sec.IV.2). We shall demonstrate that in **TLTp** one can reproduce - if a certain assumption is made - all the axioms of Tarski's [1933] metascience (Sec.IV.3). The set $\overline{S}$ is a concept which is defined separately in **DTSCL** (Sec.IV.5) and in **DTSCω-L** (Sec.IV.6). The last two dual theories are based on **TETp** (Sec.IV.4), which is turn based on **TETk** and obtained from the latter by the adding to it of definitions of all analoga of primitive and secondary concepts of **TETk**, needed to introduce the definition of $\overline{S}$ and to determine the basic properties of that set. We mean here above all the definitions of the following concepts: the *set* $\overline{I}_o$ *of basic abstract indices*, the *set* $\overline{I}$ *of all abstract indices*, the *relation* $\overline{\iota}$ *of indicating the indices of word types*, the *relation* $\overline{\rho}$ *of forming compound expression types*, and the *set* $\overline{E}$ *of all expression types*. These concepts are analoga, respectively, of the concepts of the set I_o, the set I, the function ι, the function ρ, and the set E. It is natural that the sets $\overline{I}_o$, $\overline{I}$, $\overline{E}$ are defined as the corresponding quotients of I_o, I ,E by the relation of equiformity $\approx$, and that the arguments of the relation $\overline{\iota}$ and $\overline{\rho}$ are determined by the appropriate classes of abstraction of label tokens.

Generally speaking, except for the concept of the relation $\approx$, each primitive or defined concept of a given theory, whose definitional extension yields one of the dual theories under consideration, has in that dual theory its analogon in the concept denoted by the term which differs as to its shape from the term denoting the concept in that theory by the horizontal line placed above that term. Each such concept (term) of a given dual theory will be said to be *induced* by the corresponding concept (term) of the corresponding original theory.

In the dual theory **DTSCL**, which is superstructured on **TSCL**, next to the simple categorial language $\mathscr{L}$ of expression tokens, described by the system $(\mathscr{L}_c)$ (see Sec.II.1), we will be able to consider the simple categorial language $\overline{\mathscr{L}}$ of expression types, described by the following system consisting of concepts induced by the concepts of $(\mathscr{L}_c)$:

$(\overline{\mathscr{L}}_c)$ $\quad\langle\ \overline{Lb};\ \overline{V},\ \overline{W};\ \overline{=};\ \overline{c};\ \overline{I}_o,\ \overline{I};\ \overline{\iota},\ \overline{\rho};\ \overline{E},\ \overline{S},\ \overline{Ct(S)}\ \rangle.$

Likewise, in the dual theory **DTSCω-L**, which is an extension of **TSCω-L**, next to the ω-language $\mathscr{L}$ described by the system $(\omega\text{-}\mathscr{L}_c)$ (see Sec.III.1), we shall be concerned with the categorial ω-language $\overline{\mathscr{L}}$, described syntactically by the following system consisting of concepts induced by the concepts of $(\omega\text{-}\mathscr{L}_c)$:

$(\omega\text{-}\overline{\mathscr{L}}_c)$ $\langle\overline{Lb};\ \overline{V},\ \overline{Vr},\ \overline{O},\ \overline{W};\ \overline{=};\ \overline{c};\ \overline{I}_o,\ \overline{I};\ \overline{\iota},\ \overline{\rho},\ (\overline{fv});\ \overline{E},\ \overline{S},\ \overline{Ct(S)}\rangle.$

In every dual theory the theorems which establish the properties of the concepts induced by the concepts of the initial theory pertaining to the languages of expression tokens and serving as its foundation, will hold. Those theorems will be syntactical analoga of the corresponding axioms, definitions and theorems of the initial theory, called *expressions induced* by the appropriate axioms, definitions, and theorems of the initial theory. What does it mean? For every dual theory the concept of an expression induced by a given expression can be defined in a rigorous manner. But to show the analogy between a given expression and the expression induced by it it suffices to adopt a general definition, which is not very rigorous:

DEFINITION IV.(α). *Let* **T** *be any of the theories* **TLTk**, **TETk**, **TSCL**, **TSCω-L**, *and let* **DT** *be the dual theory obtained from the latter.*

If ϕ *is an expression in* **T**, *then* $\overline{\phi}$ *is in* **DT** *an expression induced by* ϕ *if and only if* $\overline{\phi}$ *is obtained from* ϕ *by the correct replacement in the latter in every place*

(a) *of the term* $\approx$ *by the symbol* $=$,

(b) *of other constant terms of* **T** *by the terms induced by them*,

(c) *of the letter variables ranging over the sets* Lb, I, E *by the variables ranging, respectively, over the sets* $\overline{Lb}$, $\overline{I}$, $\overline{E}$,

(d) *of the variables ranging over the family* 2^{Lb} *by the variables ranging over the family* $2^{\overline{Lb}}$.[27]

An induced expression has in its verbal formulation in the dual theory a version which is quite similar to the version of the expression by which it is induced.[28]

Before we proceed to outline the dual theories **TLTp**, **TETp**, **DTSCL**, **DTSCω-L** we adopt the convention, as in case of previous chapters, stating that

C o n v e n t i o n IV.1. The theses (accepted sentences) whose numbers are marked with an asterisk, are theses of the metatheory of a given theory.

In constructing a dual theory as an extension of a given theory we shall observe the conventions applied at the presentation of that given theory.

Sec. IV.2. The theory TLTp of label types

The theory **TLTp** of label types is obtained from the theory **TLTk** of label tokens by adding to the former definitions of the concepts induced by the primitive and derived concepts of the latter.

THEOREM IV.1*. *Every thesis (accepted sentence) of* **TLTk** *is a thesis of* **TLTp**.

This theorem will be referred to on many occasions later in this chapter.

It follows from Axioms I.1a-c of **TLTk** that the relation of equiformity is an equivalence relation in the set Lb. We can accordingly produce classes of abstraction relative to that relation. The class of abstraction of the relation $\approx$, determined by the label token p (representative of that class), will be denoted by $[\ p\ ]$. As it is known, the quotient family $Lb/_{\approx}$ (the quotient of the set Lb by the relation $\approx$) is the set of all classes of abstraction of the relation $\approx$ whose representatives are definite label tokens. The set $\overline{Lb}$ of all label types is defined thus:

DEFINITION IV.1

$$\overline{Lb} = Lb/_{\approx},$$

which can be replaced by

CONCLUSION IV.1

$$\bar{p} \in \overline{Lb} \Leftrightarrow \bigvee_{p} \bar{p} = [p].$$

A label type is the class of abstraction of the relation $\approx$, determined by a certain label token.

We adopt the following convention (see Convention I.1a):

C o n v e n t i o n IV.2a. The symbols

$\bar{p}$, $\bar{q}$, $\bar{r}$, $\bar{s}$, $\bar{t}$, $\bar{u}$, $\bar{v}$, with or without subscripts,

range over the set $\overline{Lb}$.

The expressions induced by Axioms I.1a–c state, respectively, that the relation of equality = is reflexive, symmetric, and transitive in $\overline{Lb}$. We accordingly have

THEOREM IV.2*. *The expressions induced by Axioms I.1a–c of* **TLTk** *are theorems in* **TLTp**.

The three-argument relation of concatenation in $\overline{Lb}$ is defined thus:

DEFINITION IV.2

$$\overline{c}(\overline{p},\overline{q},\overline{r}) \Leftrightarrow \bigvee_{p,q,r} (\overline{p} = [p] \wedge \overline{q} = [q] \wedge \overline{r} = [r] \wedge c(p,q,r)).$$

A label type $\overline{r}$ is the concatenation of label types $\overline{p}$ and $\overline{q}$ if and only if the labels $\overline{p}$, $\overline{q}$, $\overline{r}$ are classes of abstraction of the relation $\approx$ such that a representative of the class $\overline{r}$ is the concatenation of representatives of the classes $\overline{p}$ and $\overline{q}$.

Definition IV.2 can be replaced by

CONCLUSION IV.2

$$\overline{p} = [p] \wedge \overline{q} = [q] \wedge \overline{r} = [r] \Rightarrow (\overline{c}(\overline{p},\overline{q},\overline{r}) \Leftrightarrow c(p,q,r)),$$

which follows from this definition and Theorem I.1 and Axiom I.7.

It can be demonstrated that $\overline{c}$ is a two-argument operation in $\overline{Lb}$. In fact, the expression induced by Axiom I.2, namely

THEOREM IV.1

$$\bigvee_{\overline{r}} \overline{c}(\overline{p},\overline{q},\overline{r})$$

is a theorem in **TLTp**.

If $\overline{p}$ and $\overline{q}$ are any label types, then it follows from Conclusion IV.1 that there are label tokens p_1 and q_1 such that $\overline{p} = [p_1]$ and $\overline{q} = [q_1]$. It also follows from Axiom I.2 that there is a label token r_1 such that $c(p_1,q_1,r_1)$. By Conclusion IV.2 we have that $\overline{c}(\overline{p},\overline{q},[r_1])$, from which, by Conclusion IV.1 and Convention IV.2a, we immediately arrive at Theorem IV.1. Further, if we assume that $\overline{c}(\overline{p},\overline{q},\overline{t})$, $\overline{c}(\overline{r},\overline{s},\overline{u})$ and $\overline{p}=\overline{r}$ and $\overline{q}=\overline{s}$, then it follows from Definition IV.2 that there are label tokens p_1,q_1,t_1,r_1,s_1,u_1 such that $\overline{p}=[p_1]$, $\overline{q}=[q_1]$, $\overline{t}=[t_1]$,

$\bar{r}=[r_1]$, $\bar{s}=[s_1]$, $\bar{u}=[u_1]$, $c(p_1,q_1,r_1)$, $c(r_1,s_1,u_1)$, $p_1 \approx r_1$ and $q_1 \approx s_1$. This and Axiom I.4 yield that $t_1 \approx u_1$, which in turn yields the equality $\bar{t}=\bar{u}$. Hence we have

THEOREM IV.2

$$\bar{c}(\bar{p},\bar{q},\bar{t}) \wedge \bar{c}(\bar{r},\bar{s},\bar{u}) \wedge \bar{p} = \bar{r} \wedge \bar{q} = \bar{s} \Rightarrow \bar{t} = \bar{u},$$

which is induced by Axiom I.4. From Theorems IV.1 and IV.2 we arrive at

CONCLUSION IV.2a

$$\bar{c}\colon \overline{Lb}\times\overline{Lb} \longrightarrow \overline{Lb}.$$

The relation $\bar{c}$ *is a two-argument operation in* $\overline{Lb}$.

The value of the function $\bar{c}$ for the arguments $\bar{p}$ and $\bar{q}$ can be symbolized by $\bar{p}\hat{\ }\bar{q}$ by adopting that

(*) $$\bar{r} = \bar{p}\hat{\ }\bar{q} \Leftrightarrow \bar{c}(\bar{p},\bar{q},\bar{r}).$$

We shall however abandon here this notation for the value of the function $\bar{c}$ because our intention is to demonstrate that the expressions induced by the axioms and theorems of **TLTk** are theorems in **TLTp**.

In view of Definition IV.2 and the properties of the classes of abstraction

$$[p] = [q] \Leftrightarrow p \approx q$$

we easily arrive at the following

THEOREM IV.2*b. *Expressions induced by Axioms I.3, I.5–I.8 of* **TLTk** *are theorems in* **TLTp**.

The theorem which is induced by Axiom I.5 can, with

reference to the formula (*), be recorded in the simpler form

$$(\overline{p}\,\hat{}\,\overline{q})\,\hat{}\,\overline{r} = \overline{p}\,\hat{}\,(\overline{q}\,\hat{}\,\overline{r}).$$

This theorem states that the relation of concatenation $\overline{c}$ is an associative operation in the set $\overline{Lb}$ of all label types.

The *vocabulary* $\overline{V}$ *of simple word types* is defined thus:

DEFINITION IV.3

$$\overline{V} = V/_{\approx}\,.$$

This definition can be replaced by

CONCLUSION IV.3

$$\overline{p} \in \overline{V} \Leftrightarrow \bigvee_{p \in V} \overline{p} = [p].$$

$\overline{p}$ *is a simple word type if and only if* $\overline{p}$ *is a class of abstraction of the relation of equiformity determined by a certain simple word token.*

Conclusions IV.1 and IV.3 and Axiom I.9 yield

THEOREM IV.3

$$\emptyset \neq \overline{V} \subseteq \overline{Lb},$$

which is an expression induced by Axiom I.9 of **TLTk**. We also easily obtain

THEOREM IV.2*c. *Expressions induced by Axioms I.10 and I.11 are theorems in* **TLTp**.

The set $\overline{W}$ of all word types is defined thus:

DEFINITION IV.4

$$\overline{W} = W/_{\approx}\,,$$

which can be replaced by

CONCLUSION IV.4

$$\overline{p} \in \overline{W} \Leftrightarrow \bigvee_{p \in W} \overline{p} = [p].$$

$\overline{p}$ is a word type if and only if $\overline{p}$ is a class of abstraction of the relation of equiformity determined by a certain word token.

We adopt a convention which is analogical to Convention I.1b.

Convention IV.2b. The letter

$\overline{X}$, with or without subscript,

ranges over the family $2^{\overline{Lb}}$.

The expression induced by Definition I.1a is the definition of the expression $\overline{\lambda}(\overline{X})$:

DEFINITION IV.5

$$\overline{\lambda}(\overline{X}) \Leftrightarrow \overline{V} \subseteq \overline{X} \wedge \bigwedge_{\overline{p},\overline{q},\overline{r}} (\overline{p},\overline{q} \in \overline{X} \wedge \overline{c}(\overline{p},\overline{q},\overline{r}) \Rightarrow \overline{r} \in \overline{X}).$$

By applying Conclusion IV.4, Definitions I.1a, I.1 and IV.5 we can demonstrate that

THEOREM IV.4

$$\overline{W} = \bigcap \{ \overline{X} \mid \overline{\lambda}(\overline{X}) \}.$$

This theorem is induced by Definition I.1 of the set W. The somewhat difficult proof of this theorem is given in the Annex.

It follows from Conclusions IV.4 and IV.3 that if $\overline{r} \in \overline{W} \setminus \overline{V}$, then there is an $r_1 \in W \setminus V$ such that $\overline{r} = [r_1]$. By Axiom I.12 there are then word tokens p_1 and q_1 such that $c(p_1,q_1,r_1)$. Hence, in view of Conclusions IV.2 and IV.4, $[p_1]$, $[q_1] \in \overline{W}$ and $\overline{c}([p_1],[q_1],\overline{r})$, so that we can state that $\bigvee_{\overline{p},\overline{q}\in\overline{W}} \overline{c}(\overline{p},\overline{q},\overline{r})$. We have thus proved

THEOREM IV.5

$$\overline{r} \in \overline{W}\setminus\overline{V} \Rightarrow \bigvee_{\overline{p},\overline{q}\in\overline{W}} \overline{c}(\overline{p},\overline{q},\overline{r}).$$

Every compound word type is the concatenation of a pair of word types.

Theorem IV.5 is induced by Axiom I.12.

The following theorem induced by the last axiom of **TLTk** also holds:

THEOREM IV.6

$$\overline{r} \in \overline{W} \wedge \overline{c}(\overline{p},\overline{q},\overline{r}) \Rightarrow \overline{p},\overline{q} \in \overline{W}.$$

If a word type is the concatenation of two label types, then those labels are word types.

To prove this theorem we note that $\overline{r}\in\overline{W}$ by assumption. In view of Conclusion IV.4 there is an $r_1\in W$ such that $\overline{r}=[r_1]$. Since $\overline{p},\overline{q}\in\overline{Lb}$ by Convention IV.2a, it follows from Conclusion IV.1 that there are p_1 and q_1 such that $\overline{p}=[p_1]$ and $\overline{q}=[q_1]$. This and the assumption stating that $\overline{c}(\overline{p},\overline{q},\overline{r})$ yield, by Conclusion IV.2, that $c(p_1,q_1,r_1)$. Since $r_1\in W$ we find, in view of Axiom I.13, that $p_1,q_1\in W$. In accordance with Conclusion IV.4 we have $\overline{p},\overline{q}\in\overline{W}$.

The concept of the *set* ${}^n\overline{V}$ *of all n-componential word types* is induced by the concept of the set ${}^nV$ of all n-componential word tokens. It is defined thus:

DEFINITION IV.6

$$^{n}\overline{V} = {}^{n}V/_{\approx} ,$$

which yields

CONCLUSION IV.5

$$\overline{p} \in {}^{n}\overline{V} \Leftrightarrow \bigvee_{p \in {}^{n}V} \overline{p} = [p].$$

The following expressions are induced by the inductive Definition I.2a,b of the set ^{n}V:

THEOREM IV.7

a. $\quad {}^{1}\overline{V} = \overline{V},$

b. $\quad \overline{r} \in {}^{n+1}\overline{V} \Leftrightarrow \bigvee_{\overline{p} \in {}^{n}\overline{V}} \bigvee_{\overline{q} \in \overline{V}} \overline{c}(\overline{p},\overline{q},\overline{r}).$

The easy proof of Theorem IV.7, based on Conclusions IV.5, IV.3, IV.1 and IV.2 and on Definitions I.2a,b, is left to the Reader.

The length of a given word type $\overline{p}$ will not be symbolized by $\overline{|\overline{p}|}$, as could be excepted in view of the definition of the concept induced by a given concept, formulated in Sec.IV.1, but by $|\overline{p}|$. Definition of that concept is induced by Definition I.3:

DEFINITION IV.7

$$\overline{p} \in \overline{W} \Rightarrow (|\overline{p}| = n \Leftrightarrow \overline{p} \in {}^{n}\overline{V}).$$

The generalized definition of the relation of concatenation $\overline{c}^{k}$ of k+1 arguments ($k \geq 2$), which holds between label types, is defined thus:

DEFINITION IV.8

$$k > 1 \Rightarrow (\overline{c}^{k}(\overline{p}_1,\overline{p}_2,\ldots,\overline{p}_k,\overline{p}) \Leftrightarrow \bigvee_{p_1,p_2,\ldots,p_k,p} (\overline{p}_1 = [p_1] \wedge$$

$$\wedge\ \overline{p}_2 = [p_2] \wedge \ldots \wedge \overline{p}_k = [p_k] \wedge \overline{p} = [p] \wedge c^{k}(p_1,p_2,\ldots,p_k,p))).$$

The following two theorems are induced by the inductive Definition I.4a,b of the generalized relation of concatenation holding between label tokens:

THEOREM IV.8

a. $n = 1 \Rightarrow (\overline{c}^{n+1}(\overline{p}_1,\overline{p}_2,\overline{p}) \Leftrightarrow \overline{c}(\overline{p}_1,\overline{p}_2,\overline{p})),$

b. $n > 1 \Rightarrow (\overline{c}^{n+1}(\overline{p}_1,\overline{p}_2,\ldots,\overline{p}_{n+1},\overline{p}) \Leftrightarrow$

$$\Leftrightarrow \bigvee_{q} (\overline{c}^{n}(\overline{p}_1,\overline{p}_2,\ldots,\overline{p}_n,\overline{q}) \wedge \overline{c}(\overline{q},\overline{p}_{n+1},\overline{p}))).$$

Theorem IV.8a follows directly from Definitions IV.8, I.4a and IV.2. Theorem IV.8b follows from Definitions IV.8 and I.4b, Convention IV.2a, Conclusion IV.1, Definition IV.2, Theorem I.1 and Axiom I.1a.

Note also that Definition IV.8 and Conclusion I.11 yield

CONCLUSION IV.6

$$k > 1 \wedge \overline{p}_1 = [p_1] \wedge \overline{p}_2 = [p_2] \wedge \ldots \wedge \overline{p}_k = [p_k] \wedge \overline{p} = [p] \Rightarrow$$

$$\Rightarrow (\overline{c}^{k}(\overline{p}_1,\overline{p}_2,\ldots,\overline{p}_k,\overline{p}) \Leftrightarrow c^{k}(p_1,p_2,\ldots,p_k,p)).$$

This conclusion can replace Definition IV.8.

We have demonstrated that in the dual theory **TLTp** every expression induced by an axiom of **TLTk** is a theorem in **TLTp** (see Theorems IV.2*a, IV.1, IV.2, IV.2*b, IV.3, IV.2*c, IV.5, IV.6). Moreover, every definition in **TLTk** has in **TLTp** its analogon in the form of an expression induced by it, such an expression being a definition or a theorem in **TLTp** (see Definition IV.5, Theorems IV.4 and IV.7a,b, Definition IV.7 and Theorem IV.8). Thus we obtain

THEOREM IV.2*. *Every expression induced by an axiom or a definition in* **TLTk** *is a thesis in* **TLTp**.

This theorem implies

THEOREM IV.3*. *Every expression induced by a thesis in* **TLTk** *is a thesis in* **TLTp**.

Sec. IV.3. Interpretation in TLTp of Tarski's axioms of metascience

A.Tarski [1933] gave a system of specific axioms which suffices to found metascience. The following names of signs and expressions of the language of the algebra of classes are the primitive terms occurring in the notation of those axioms: "sign of negation", "sign of logical sum", "universal quantifier", "sign of inclusion", "variable of k-th shape", "expression consisting of two successive expressions x and y", and "expression". Tarski used, respectively, the symbols: ng, sm, qu, in, v_k, and $x^\frown y$, as abbreviations of those terms except for the last one. In recording the axioms of metascience Tarski also assumed that the variables x,y,z,t,u,w range over the set of expressions (and sequences of expressions) treated as classes of concrete labels which are equiform with a given label, and that the variable X ranges over the class of such expressions.

We quote here Tarski's axioms of metascience:

AXIOM 1. *ng*, *sm*, *qu* and *in* are expressions; no two expressions out of these four are identical.

AXIOM 2. v_k is an expression if and only if k is a natural number other than 0; v_k differs from the expressions *ng*, *sm*, *qu* and *in* and from every expression v_1 if $k \neq 1$.

AXIOM 3. $x^\frown y$ is an expression if and only if x and y are expressions; $x^\frown y$ differs from the expressions *ng*, *sm*, *qu* and *in* and from every expression v_k.

AXIOM 4. If x, y, z and t are expressions, then $x^\frown y = z^\frown t$ if and only if one of the following conditions is satisfied: (α) $x=z$ and $y=t$; (β) there is an expression u such that $x=z^\frown u$ and $t=u^\frown y$; (γ) there is an expression u such that $z=x^\frown u$ and $y=u^\frown t$.

AXIOM 5 (principle of mathematical induction). If a class X satisfies the following conditions: (α) $ng \in X$, $sm \in X$, $qu \in X$ and $in \in X$; (β) if k is a natural number other than 0, then $v_k \in X$; (γ) if $x \in X$ and $y \in X$, then $x^\frown y \in X$, then every expression is in the class X.

Tarski's axioms of metascience can be found in **TLTp** (cf. G.Bryll and S.Miklos [1977]) by enriching the language of the latter by the terms *ng*, *sm*, *qu*, *in*, v_k ($k \in N$) and by assuming that the thesis

(**) $$|\overline{V}| = \aleph_0,$$

which states that the vocabulary of simple word types is an infinite enumerable set, is valid in **TLTp**. We can arrive at this thesis if we postulate in **TLTk** that there are infinitely enumerably many diversiform simple word tokens.

In such a case we obtain an interpretation in **TLTp** of Tarski's axiom system given above by setting $\{ng, sm, qu, in, v_1, v_2, \ldots\} = \overline{V}$ and by replacing the variables $x, y, z, t, u,$ respectively by the variables $\overline{p}, \overline{q}, \overline{r}, \overline{t}, \overline{u},$ the variable X by the variable $\overline{X}$, the expression $z = x\widehat{\ }y$ by the expression $\overline{r} = \overline{p}\widehat{\ }\overline{q}$, i.e., by the expression $\overline{c}(\overline{p}, \overline{q}, \overline{r})$ (see formula (*)), and the term "expression" by the term "word type".

Since, by Theorem IV.3*, those theorems which are expressions induced by Conclusion I.3a and Theorems I.6 and I.4a are valid in **TLTp**, we arrive at

CONCLUSION IV.7

$$\overline{V} \subseteq \overline{W},$$

CONCLUSION IV.8

$$\overline{c}(\overline{p}, \overline{q}, \overline{r}) \Rightarrow (\overline{r} \in \overline{W} \Leftrightarrow \overline{p}, \overline{q} \in \overline{W}),$$

CONCLUSION IV.9

$$\overline{r} \in \overline{W} \setminus \overline{V} \Leftrightarrow \bigvee_{\overline{p}, \overline{q} \in \overline{W}} \overline{c}(\overline{p}, \overline{q}, \overline{r}).$$

We can easily find that for the above interpretation Tarski's Axioms 1-3 become theses in **TLTp**.

The expression

$$\overline{c}(\overline{p},\overline{q},\overline{s}) \wedge \overline{c}(\overline{r},\overline{t},\overline{v}) \Rightarrow (\overline{s} = \overline{v} \Leftrightarrow \overline{p} = \overline{r} \wedge \overline{q} = \overline{t} \vee$$

$$\vee \bigvee_{u} (\overline{c}(\overline{r},\overline{u},\overline{p}) \wedge \overline{c}(\overline{u},\overline{q},\overline{t}) \vee \overline{c}(\overline{p},\overline{u},\overline{r}) \wedge \overline{c}(\overline{u},\overline{t},\overline{q})))$$

is also a theorem in **TLTp**, induced by Theorem I.2. From this expression, by Conclusion IV.8, we arrive at

CONCLUSION IV.10

$$\overline{p},\overline{q},\overline{r},\overline{t} \in \overline{W} \wedge \overline{c}(\overline{p},\overline{q},\overline{s}) \wedge \overline{c}(\overline{r},\overline{t},\overline{v}) \Rightarrow (\overline{s} = \overline{v} \Leftrightarrow \overline{p} = \overline{r} \wedge \overline{q} = \overline{t} \vee$$

$$\vee \bigvee_{\overline{u}\in\overline{W}} (\overline{c}(\overline{r},\overline{u},\overline{p}) \wedge \overline{c}(\overline{u},\overline{q},\overline{t})) \vee \bigvee_{\overline{u}\in\overline{W}} (\overline{c}(\overline{p},\overline{u},\overline{r}) \wedge \overline{c}(\overline{u},\overline{t},\overline{q}))).$$

The above conclusion, after the application of the formula (*), is an expression into which Tarski's Axiom 4 turns for the interpretation given above.

We have to demonstrate that, for the interpretation given above, Tarski's Axiom 5 becomes a theorem in **TLTp**. It is so in fact because, by Theorem IV.3*, the conclusion induced by Conclusion I.3c, that is

CONCLUSION IV.11

$$\overline{V} \subseteq \overline{X} \wedge \bigwedge_{\overline{p},\overline{q},\overline{r}} (\overline{p},\overline{q} \in \overline{X} \wedge \overline{c}(\overline{p},\overline{q},\overline{r}) \Rightarrow \overline{r} \in \overline{X}) \Rightarrow \overline{W} \subseteq \overline{X},$$

is a thesis in **TLTp**.

Sec. IV.4. The theory TETp of expression types

The theory **TETp** of expression types is superstructured on the theory **TETk** of expression tokens, and hence on the theory **TLTk** of label tokens.

We note accordingly

THEOREM IV.4*. *Every thesis (accepted sentence) of* **TETk**, *and in particular of* **TLTk**, *is a thesis in* **TETp**.

Now **TETp** is obtained from **TETk** by the joining to the latter of all concepts induced by the primitive and derived concepts of **TETk**. The concepts induced by the primitive and derived concepts of **TLTk**, which is a fragment of **TETk**, have already been defined in **TLTp** (Sec.IV.2). It can thus be seen that **TETp** is based on **TLTp** and the following theorem holds:

THEOREM IV.5*. *Every thesis (accepted sentence) of* **TLTp** *is a thesis in* **TETp**.

Theorems IV.4* and IV.5* will hereafter be used tacitly.

Our task now reduces to defining the concepts induced by those primitive and derived concepts of **TETk** which are not concepts in **TLTk** and to formulating the basic theorems which describe the properties of the concepts so introduced.

The *concept of the set* $\overline{I}_o$ *of all basic abstract indices* - induced by the concept of the set I_o of all basic indices - is defined thus:

DEFINITION IV.9

$$\overline{I}_o = I_o/_{\approx},$$

which can be replaced by

CONCLUSION IV.12

$$\overline{p} \in \overline{I}_o \Leftrightarrow \bigvee_{p \in I_o} \overline{p} = [p].$$

A basic abstract index is the class of abstraction consisting of all indices equiform with a certain basic index.

The *set* $\overline{I}$ *of all abstract indices* - the induced by the concept of the set I of all indices - is defined thus:

DEFINITION IV.10

$$\overline{I} = I/_{\approx} ,$$

which can be replaced by

CONCLUSION IV.13

$$\overline{p} \in \overline{I} \Leftrightarrow \bigvee_{p \in I} \overline{p} = [p].$$

An abstract index is the class of abstraction consisting of all indices equiform with a certain index.

By making use of

DEFINITION IV.11

$$\overline{\lambda}(\overline{X})^{d} \Leftrightarrow \overline{I}_{o} \subseteq \overline{X} \wedge \bigwedge_{\overline{p},\overline{q},\overline{r}} (\overline{p},\overline{q} \in \overline{X} \wedge \overline{c}(\overline{p},\overline{q},\overline{r}) \Rightarrow \overline{r} \in \overline{X}),$$

which is induced by Definition II.1a, we can prove

THEOREM IV.9

$$\overline{I} = \bigcap \{ \overline{X} \mid \overline{\lambda}(\overline{X})^{d} \},$$

induced by Definition II.1 of the theory **TETk**. Its proof, analogical to the proof of Theorem IV.4, is omitted.

The basic properties of the sets $\overline{I}_{o}$ and $\overline{I}$ are established by theorems induced by Axioms II.1 and II.3-II.5:

THEOREM IV.10

$$\emptyset \neq \overline{I}_{o} \subseteq \overline{Lb};$$

THEOREM IV.11

$$\overline{c}(\overline{p},\overline{q},\overline{r}) \Rightarrow \overline{r} \notin \overline{I}_{o};$$

THEOREM IV.12

$$\bar{r} \in \bar{I} \setminus \bar{I}_o \Rightarrow \bigvee_{\bar{p},\bar{q},\bar{r}} \bar{c}(\bar{p},\bar{q},\bar{r});$$

THEOREM IV.13

$$\bar{r} \in \bar{I} \wedge \bar{c}(\bar{p},\bar{q},\bar{r}) \Rightarrow \bar{p},\bar{q} \in \bar{I}.$$

Theorem IV.10 follows immediately from Conclusion IV.12, Axiom II.1 and Conclusion IV.1. Theorem IV.11 follows from Definition IV.2, Axioms II.3 and II.2 and Conclusion IV.12. Theorem IV.12 follows from Conclusions IV.13 and IV.12, Axiom II.4 and Definition IV.2. Theorem IV.13 follows directly from Conclusion IV.13, Definition IV.2, Axioms I.7 and II.5, and Conclusion IV.13 again.

We self-evidently have the formula

$$(***) \qquad \bar{p} \in \bar{I}_o \wedge \bar{q} = \bar{p} \Rightarrow \bar{q} \in \bar{I}_o,$$

which is induced by Axiom II.2 of TETk.

The concept of the *set* $^n\bar{I}$ *of all n-componential abstract indices* ($n \geq 1$) - induced by the concept of the set ^nI - is defined thus:

DEFINITION IV.12

$$^n\bar{I} = {}^nI/_{\approx}\,.$$

This definition is equivalent to

CONCLUSION IV.14

$$\bar{p} \in {}^n\bar{I} \Leftrightarrow \bigvee_{p \in {}^nI} \bar{p} = [p].$$

The following theorems are induced by Definitions II.2a,b:

THEOREM IV.14

a. $\quad {}^{1}\bar{I} = \bar{I}_{o}$,

b. $\quad \bar{r} \in {}^{n+1}\bar{I} \Leftrightarrow \bigvee_{\bar{p}\in{}^{n}\bar{I}} \bigvee_{\bar{q}\in\bar{I}_{o}} \bar{c}(\bar{p},\bar{q},\bar{r})$.

The easy proofs of these theorems, based on Conclusions IV.14 and IV.12, Definitions II.2a,b, Conclusions IV.2 and IV.1, and Convention IV.2a, are omitted.

Note now that Remarks II.2 and II.3, Definition IV.(α), Theorems IV.2*a,b, IV.1 and IV.2, Definition IV.11, Theorems IV.8–IV.14, and formula (***) yield

THEOREM IV.6*. *The expressions induced by the dual analoga of all axioms and definitions adopted in* **TLTk** *are theses in* **TETp**.[29]

Theorem IV.6* implies

THEOREM IV.7*. *Every expression induced by the dual analogon of a thesis in* **TLTk** *is a thesis in* **TETp**.

In view of Theorem IV.7* we can see, for instance, that the expression $\bar{I}_{o} \subseteq \bar{I} \subseteq \overline{Lb}$, induced by Conclusion I.3^{d}a (dual analogon of Conclusion I.3a), is a theorem in **TETp**.

Note further that the theorem induced by Axiom II.6 in **TETk** holds in **TETp**:

THEOREM IV.15

$$\bar{V} \cap \bar{I}_{o} = \emptyset.$$

The vocabulary of a simple word types has no elements in common with the set of all basic abstract indices.

Theorem IV.15 follows easily from Conclusions IV.3 and IV.12 and Axioms I.10 and II.6.

The relation ι of indicating the indices of word tokens is a primitive concept in **TETk**. The concept of the *relation* $\bar{\iota}$ *of indicating the indices of word types*, induced by the former, is defined thus:

DEFINITION IV.13

$$\bar{\iota}(\bar{p},\bar{q}) \Leftrightarrow \bigvee_{p\in D(\iota)} (\bar{p} = [p] \wedge \bar{q} = [\iota(p)]).$$

Definition IV.13 and Axiom II.7 yield immediately

CONCLUSION IV.15

a. $\bar{p} \in D(\bar{\iota}) \Leftrightarrow \bigvee_{p\in D(\iota)} \bar{p} = [p],$

b. $\bar{q} \in D^{-1}(\bar{\iota}) \Leftrightarrow \bigvee_{q\in D^{-1}(\iota)} \bar{q} = [q].$

The relation $\bar{\iota}$ has as its domain the set of all classes of abstraction of the relation $\approx$ determined by those word tokens which have an index. It has as its counter-domain the set of all classes of abstraction of the relation $\approx$ determined by the indices of word tokens.

It follows from the definition of $\bar{\iota}$ and from Conclusion II.2b that $\bar{\iota}$ is a function. Since it follows from Conclusions IV.15a, II.1a and IV.4 that its domain is a subset of $\bar{W}$, and from Conclusions IV.15b, II.1b and IV.13, that its counter-domain is a subset of $\bar{I}$, we obtain

THEOREM IV.16

$$\bar{\iota} \subseteq \bar{W}\times\bar{I} \wedge \bar{\iota} \text{ is a function.}$$

This theorem is induced by Axiom II.7 of **TETk**.

On denoting by $\bar{\iota}(\bar{p})$ the value of $\bar{\iota}$ for the argument $\bar{p}$ of its domain we obtain, by Definition IV.13 and Conclusions IV.15a and II.2b, the following

THEOREM IV.16a

$$\bar{p} = [p] \wedge p \in D(\iota) \Rightarrow \bar{\iota}(\bar{p}) = [\iota(p)].$$

It is self-evident that

THEOREM IV.8*a. *The expression induced by Axiom II.8 of* **TETk** *is a theorem in* **TETp**.

The operation ρ of forming compound expression tokens has its analogon in the induced concept of the binary *relation* $\bar{\rho}$ *of forming compound expression types*. The elements of the domain of that relation are finite ordered sequences of definite word types, and the elements of its counter-domain are definite word types. In recording the definition of the relation $\bar{\rho}$ we shall use the expression $\bar{\rho}(\bar{p}_0,\bar{p}_1,\dots,\bar{p}_n;\bar{p})$, to be read: the relation $\bar{\rho}$ holds between the sequence of words $(\bar{p}_0,\bar{p}_1,\dots,\bar{p}_n)$ and the word $\bar{p}$, or: $\bar{p}$ is a compound expression type consisting of the word types $p_0,p_1,\dots,p_n$, in that order.

DEFINITION IV.14

$$n \geq 1 \Rightarrow \{ \bar{\rho}(\bar{p}_0,\bar{p}_1,\dots,\bar{p}_n;\bar{p}) \Leftrightarrow \bigvee_{t_0,t_1,\dots,t_n \in D(\iota)} (\bar{p}_0 = [t_0] \wedge$$

$$\wedge \bar{p}_1 = [t_1] \wedge \dots \wedge \bar{p}_n = [t_n] \wedge \bar{p} = [\rho(t_0,t_1,\dots,t_n)]) \}.$$

It follows from this definition and Conclusion II.3 that $\bar{\rho}$ is a one-to-one function. Further it follows from Definition IV.14, Conclusion IV.15a, Axiom II.9, Conclusion IV.3 and Axiom

I.10 that that function has the set $\bigcup_{k=2}^{\infty} D(\bar{\iota})^k$ as its domain and a subset of $D(\bar{\iota})\setminus\bar{V}$ as its counter-domain. This yields the following theorem, induced by Axiom II.9 of **TETk**:

THEOREM IV.17

$$\bar{\rho}\colon \bigcup_{k=2}^{\infty} D(\bar{\iota})^k \xrightarrow[1-1]{} D(\bar{\iota})\setminus\bar{V}.$$

On denoting by $\bar{\rho}(\bar{p}_0,\bar{p}_1,\dots,\bar{p}_n)$ the value of $\bar{\rho}$ for the argument $(\bar{p}_0,\bar{p}_1,\dots,\bar{p}_n)$ of its domain we obtain, by Definition IV.14 and Conclusions IV.15 and II.3,

THEOREM IV.17a

$$n \geq 1 \wedge t_0,t_1,\dots,t_n \in D(\iota) \wedge \bar{p}_0 = [t_0] \wedge \bar{p}_1 = [t_1] \wedge \dots \wedge$$
$$\wedge\ \bar{p}_n = [t_n] \Rightarrow \bar{\rho}(\bar{p}_0,\bar{p}_1,\dots,\bar{p}_n) = [\rho(t_0,t_1,\dots,t_n)].$$

A compound expression type - consisting, successively, of word types $\bar{p}_0,\bar{p}_1,\dots,\bar{p}_n$, determined by classes of abstraction of the relation $\approx$, represented, respectively, by indexed word tokens $t_0,t_1,\dots,t_n$ - is a class of abstraction of the relation $\approx$, represented by compound expression token consisting of the words $t_0,t_1,\dots,t_n$, in that order, that is of the functor t_0 and its successive arguments $t_1,\dots,t_n$.

Self-evidently we have

THEOREM IV.8*b. *The expression induced by Axiom II.10 of* **TETk** *is a theorem in* **TETp**.

We pass now to definitions of concepts induced by the concepts of the sets E_s, E_c and E. The *set* $\bar{E}_s$ *of all simple expression types* is defined thus:

DEFINITION IV.15

$$\overline{E}_S = E_S/_{\approx} .$$

This definition can be replaced by

CONCLUSION IV.16

$$\overline{\rho} \in \overline{E}_S \Leftrightarrow \bigvee_{\rho \in E_S} \overline{\rho} = [\rho].$$

For the *set* $\overline{E}_C$ *of all compound expression types* we have

DEFINITION IV.16

$$\overline{E}_C = E_C/_{\approx} ,$$

which can be replaced by

CONCLUSION IV.17

$$\overline{\rho} \in \overline{E}_C \Leftrightarrow \bigvee_{\rho \in E_C} \overline{\rho} = [\rho].$$

For the *set* $\overline{E}$ *of all expression types* we have naturally

DEFINITION IV.17

$$\overline{E} = E/_{\approx} ,$$

which can be replaced by

CONCLUSION IV.18

$$\overline{\rho} \in \overline{E} \Leftrightarrow \bigvee_{\rho \in E} \overline{\rho} = [\rho].$$

Conclusion IV.16, Definition II.3a, Conclusions IV.3 and IV.15a, and Axiom I.10 yield

THEOREM IV.18a

$$\overline{E}_S = \overline{V} \cap D(\overline{\iota}).$$

Conclusion IV.17, Definition II.3b, Axiom II.9, Theorem IV.17a, Conclusion IV.15a and Theorem IV.17 yield

THEOREM IV.18b

$$\overline{E}_{\mathrm{C}} = D^{-1}(\overline{\rho}).$$

Finally, by Conclusion IV.18, Definition II.3c and Conclusions IV.16 and IV.17, we have

THEOREM IV.18c

$$\overline{E} = \overline{E}_{\mathrm{S}} \cup \overline{E}_{\mathrm{C}}.$$

Theorems IV.18a,b,c are induced, respectively, by Definitions II.3a,b,c.

We now adopt conventions analogical to Conventions II.3a-c. We shall henceforth have to observe, next to Conventions II.3a-c, the following

C o n v e n t i o n IV.3a. The variables

$\overline{p},\overline{q},\overline{r},\overline{s}$, with or without subscripts,

range over the set $\overline{E}$.

b. The variables

$\overline{a},\overline{b},\overline{c},\overline{d},\overline{\xi},\overline{\sigma}$, with or without subscripts,

range over the set $\overline{I}$.

c. $\overline{\iota}(\overline{p}) = \overline{a} \wedge \overline{\iota}(\overline{q}) = \overline{b} \wedge \overline{\iota}(\overline{r}) = \overline{c} \wedge \overline{\iota}(\overline{s}) = \overline{d},$

$\bigwedge_{i \in N \cup \{0\}} (\overline{\iota}(\overline{p}_i) = \overline{a}_i \wedge \overline{\iota}(\overline{q}_i) = \overline{b}_i \wedge \overline{\iota}(\overline{r}_i) = \overline{c}_i \wedge \overline{\iota}(\overline{s}_i) = \overline{d}_i.$

The *syntactic category* $\overline{Ct}_{\overline{\xi}}$ *with the index* $\overline{\xi}$ is defined by the formula

DEFINITION IV.18

$$\overline{p} \in \overline{Ct}_{\overline{\xi}} \Leftrightarrow \bigvee_{\xi} \bigvee_{p \in Ct_{\xi}} (\overline{p} = [p] \wedge \overline{\xi} = [\xi]).$$

An expression type $\overline{p}$ is in the syntactic category with the abstract index $\overline{\xi}$ if and only if $\overline{p}$ is a class of abstraction of the relation $\approx$ which has as its representative an expression token such that it is in the syntactic category with the index which is a representative of that class of abstraction of the relation $\approx$ which determines the index $\overline{\xi}$.

From the above definition, Definition II.8, Conventions II.3a-c and IV.3a-c, Conclusion II.5b, Theorem IV.16a, and Conclusions IV.18 and IV.13 we obtain

THEOREM IV.19

$$\overline{Ct}_{\overline{\xi}} = \{ \overline{p} \mid \overline{\alpha} = \overline{\xi} \}.$$

The syntactic category with the index $\overline{\xi}$ is the set of all expression types whose index is equal to the index $\overline{\xi}$.

Note that Theorem IV.19 is induced by Definition II.8.

The concept induced by the family Ct of all syntactic categories of expression tokens is that of the *family $\overline{Ct}$ of all syntactic categories of expression types*, defined as follows.

DEFINITION IV.19

$$\overline{Ct} = \{ \overline{Ct}_{\overline{\xi}} \mid \overline{\xi} \in \overline{\iota}(\overline{E}) \}.$$

The relation $\underset{c}{=}$ of syntactic categorial agreement of expression tokens has its analogon in the *relation $\underset{c}{\equiv}$ of syntactic categorial agreement of expression types* described by the following definition induced by Definition II.10:

DEFINITION IV.20

$$\overline{p} \underset{c}{\equiv} \overline{q} \Leftrightarrow \bigvee_{\overline{\xi}} \overline{p},\overline{q} \in \overline{Ct}_{\overline{\xi}}.$$

Two expression types $\overline{p}$ and $\overline{q}$ belong to the same syntactic category if they belong to a syntactic category with the same abstract index.

When formulating the theory **TETk** we have adopted three other concepts: that of the set C^n_p of all constituents of the n-th order of an expression token p, that of the set C_p of all constituents of an expression token p, and that of the set $C_p^{(j_1,j_2,\ldots,j_n)}$ of those constituents of an expression token p which occupy in it the syntactic positions $(j_1,j_2,\ldots,j_n)$ $(n>0)$. The concepts induced by these concepts will be denoted, respectively, by $\overline{C}^n_{\overline{p}}$, $\overline{C}_{\overline{p}}$ and $\overline{C}_{\overline{p}}^{(j_1,j_2,\ldots,j_n)}$. The definitions of those concepts are as follows.

DEFINITION IV.21a

$$\overline{t} \in \overline{C}^n_{\overline{p}} \Leftrightarrow \bigvee_{t,p} (\overline{t} = [t] \wedge \overline{p} = [p] \wedge t \in C^n_p).$$

DEFINITION IV.21

$$\overline{t} \in \overline{C}_{\overline{p}} \Leftrightarrow \bigvee_{t,p} (\overline{t} = [t] \wedge \overline{p} = [p] \wedge t \in C_p).$$

DEFINITION IV.22

$$\overline{t} \in \overline{C}_{\overline{p}}^{(j_1,j_2,\ldots,j_n)} \Leftrightarrow \bigvee_{t,p} (\overline{t} = [t] \wedge \overline{p} = [p] \wedge$$
$$\wedge\ t \in C_p^{(j_1,j_2,\ldots,j_n)}).$$

It can be proved, by reference to Definitions IV.21a and II.12a–c, Conclusions IV.18 and IV.1, Theorem IV.17a and

Conclusions IV.15a and II.16a, that the expression induced by Definitions II.12a-c are theorems in **TETp**. Hence we have

THEOREM IV.20

a. $\bar{t} \in \bar{C}^{0}_{\bar{\rho}} \Leftrightarrow \bar{\rho} = \bar{t},$

b. $\bar{t} \in \bar{C}^{1}_{\bar{\rho}} \Leftrightarrow \bigvee_{\bar{t}_0,\bar{t}_1,\ldots,\bar{t}_n \in D(\bar{\iota})} (\bar{\rho} = \bar{\rho}(\bar{t}_0,\bar{t}_1,\ldots,\bar{t}_n) \wedge$

$\wedge \bigvee_{0\le j\le n} \bar{t} = \bar{t}_j),$

c. $k > 0 \Rightarrow (\bar{t} \in \bar{C}^{k+1}_{\bar{\rho}} \Leftrightarrow \bigvee_{\bar{r}\in\bar{C}^{k}_{\bar{\rho}}} \bar{t} \in \bar{C}^{1}_{\bar{r}}).$

The following expression induced by Definition II.12d follows immediately from Definitions IV.21, II.12d and IV.21a:

THEOREM IV.20d

$$\bar{t} \in \bar{C}_{\bar{\rho}} \Leftrightarrow \bigvee_{n} \bar{t} \in \bar{C}^{n}_{\bar{\rho}} .$$

By making use, in particular, of Definitions IV.22 and II.13a,b and Theorem IV.17a we can prove

THEOREM IV.21

a. $\bar{t} \in \bar{C}^{(j_1)}_{\bar{\rho}} \Leftrightarrow \bar{\rho} \in \bar{E}_{\mathrm{C}}$ *and* $\bar{t}$ *is the* j_1*-th word of the expansion of* $\bar{\rho}$,

b. $\bar{t} \in \bar{C}^{(j_1,j_2,\ldots,j_{k+1})}_{\bar{\rho}} \Leftrightarrow \bar{t}$ *is the* j_{k+1}*-th word of the expansion of an element of the set* $\bar{C}^{(j_1,j_2,\ldots,j_k)}_{\rho}$ *for* $k>0$.[30]

The last theorem gives expressions induced by Definition II.13a,b.

It is self-evident that every axiom and every definition of **TETk** has its analogon in an induced expression which is a thesis in **TETp**. The fact that the expressions induced by Axioms II.1–II.10 of **TETk** are theorems in **TETp** can be established on the basis of Theorems IV.10–IV.13, formula (***), and Theorems IV.15, IV.16, IV.8*a, IV.17, and IV.8*b. Definition IV.11, Theorems IV.9, IV.14a,b, IV.18a–c, IV.19, Definitions IV.19 and IV.20, and Theorems IV.20a–d and IV.21a,b are induced by Definitions II.1a, II.1, II.2a,b, II.3a–c, II.8, II.9, II.10, II.12a–d, and II.13a,b, respectively. This substantiates

THEOREM IV.8*. *Every expression induced by an axiom or a definition of* **TETk** *is a thesis in* **TETp**.

Theorem IV.8* implies

THEOREM IV.9*. *Every expression induced by a thesis of* **TETk** *is a thesis in* **TETp**.

It follows from Theorem IV.9* that if $E \neq \emptyset$ then we have, in particular, a theorem induced by Theorem II.6. It states that the family $\overline{Ct}$ of all syntactic categories of expression types is a logical classification of the set $\overline{E}$, determined by the equivalence relation $\underset{c}{\equiv}$.

Sec. IV.5. The dual theory DTSCL

The theory **TSCL** of simple categorial languages of expression tokens is obtained from the theory **TETk** by the joining to the latter of one axiom (Axiom II.11) and several definitions, including the definition of the set S of all well-formed expression tokens of a given simple categorial language $\mathscr{L}$. The dual theory **DTSCL**, which makes it possible to analyse the language $\overline{\mathscr{L}}$ described by the system $(\overline{\mathscr{L}}_c)$ (see Sec.IV.1), is obtained from **TSCL** by joining to the latter the definitions of all concepts induced by the primitive and derived concepts of **TSCL**. The concepts induced by the primitive and derived concepts of **TETk** have already been defined in **TETp**. It follows therefore from the above that **DTSCL** is not only superstructured on **TSCL** but on **TETp** as well. We accordingly have

THEOREM IV.10*. *Every thesis (accepted sentence) of* **TSCL**, *and in particular of* **TETk**, *is a thesis in* **DTSCL**.

and

THEOREM IV.11*. *Every thesis (accepted sentence) of* **TETp** *is a thesis in* **DTSCL**.

We shall henceforth make tacit references to Theorems IV.10* and IV.11*.

The theory **DTSCL** includes, above all, the following definition of the *set* $\overline{S}$ *of all well-formed expression types* of the language $\overline{\mathscr{L}}$:

DEFINITION IV.23

$$\overline{S} = S/_{\approx}\,,$$

which can be replaced by

CONCLUSION IV.19

$$\overline{p} \in \overline{S} \Leftrightarrow \bigvee_{p \in S} \overline{p} = [p].$$

A well-formed expression type of the language $\overline{\mathscr{L}}$ is a class of abstraction consisting of all expressions equiform with a certain well-formed expression token of $\mathcal{L}$.

The *set* $^{n}\overline{S}$ *of all well-formed expression types of order* n ($n \geq 0$) - a concept induced by the concept of the set ^{n}S introduced in **TSCL** - is defined as a quotient family: the quotient of the set ^{n}S by the relation of equiformity. Hence we have

Definition IV.23a

$$^{n}\overline{S} = {}^{n}S/_{\approx} .$$

This definition can be replaced by the following

Conclusion IV.20

$$\overline{p} \in {}^{n}\overline{S} \Leftrightarrow \bigvee_{p \in {}^{n}S} \overline{p} = [p].$$

By making use of the definitions adopted above we have first of all to demonstrate that the expressions induced by the inductive Definition II.4a,b of the set ^{n}S and the expression induced by the Definition IV.4c of the set S are theorems in **DTSCL**.

Theorem IV.22

a. $$^{0}\overline{S} = \overline{E}_{s},$$

b. $$\overline{p} \in {}^{k+1}\overline{S} \Leftrightarrow \overline{p} \in {}^{k}\overline{S} \vee \bigvee_{n \geq 1} \bigvee_{\overline{p}_0, \overline{p}_1, \ldots, \overline{p}_n \in {}^{k}\overline{S}} (\overline{p} = \overline{\rho}(\overline{p}_0, \overline{p}_1, \ldots, \overline{p}_n) \wedge \overline{c}^{n+1}(\overline{\alpha}, \overline{\alpha}_1, \ldots, \overline{\alpha}_n, \overline{\alpha}_0)),$$

c. $$\overline{S} = \bigcup_{n=0}^{\infty} {}^{n}\overline{S}.$$

Theorem IV.22a follows immediately from Conclusions IV.20 and IV.16 and Definition II.4a. The somewhat more difficult proof of Theorem IV.22b is given in the Annex. Theorem IV.22c follows easily from Conclusion IV.19, Definition II.4c and Conclusion IV.20.

In the theory **TSCL** we have adopted an auxiliary definition of the condition $\delta(X)$ and we have proved Theorem II.2 which states that the set S is the least set which satisfies that condition. In **DTSCL** the expression $\overline{\delta}(\overline{X})$ is induced by $\delta(X)$. It can be shown that in the latter theory a theorem analogical to Theorem II.2 is valid. Note that

REMARK IV.1. In **DTSCL** an expression induced by the Definition II.5 of the condition $\delta(X)$ is the definition of the condition $\overline{\delta}(\overline{X})$.

The following theorem induced by Axiom II.11 is also a theorem in **DTSCL**.

THEOREM IV.23

$$\overline{\iota}(\overline{S}\setminus{}^{0}\overline{S}) \cap \overline{I}_{0} \neq \emptyset.$$

Theorem IV.23 follows easily from Axiom II.11, Definition II.4a, Conclusions IV.19, IV.20, II.4a, IV.12 and II.6b and Theorem IV.16a.

The *set* $\overline{B}$ *of all basic expression types* of $\overline{\mathcal{L}}$ and the *set* $\overline{F}$ *of all functor types* of $\overline{\mathcal{L}}$ are concepts induced by the concepts of the set B of all expressions of $\mathcal{L}$ and the set F of all functors of $\mathcal{L}$, respectively. These concepts self-evidently are defined thus:

DEFINITION IV.24

$$\overline{B} = B/_{\approx} ,$$

and

DEFINITION IV.25

$$\overline{F} = F/_{\approx} .$$

We use the resulting conclusions

CONCLUSION IV.21

$$\overline{p} \in \overline{B} \Leftrightarrow \bigvee_{p \in B,} \overline{p} = [p]$$

and

CONCLUSION IV.22

$$\overline{p} \in \overline{F} \Leftrightarrow \bigvee_{p \in F} \overline{p} = [p]$$

to formulate theorems which are, respectively, induced by Definitions II.6 and II.7, adopted in **TSCL**:

THEOREM IV.24

$$\overline{B} = \{ \overline{p} \in \overline{S} \mid \overline{\alpha} \in \overline{I}_{o} \}$$

and

THEOREM IV.25

$$\overline{F} = \{ \overline{p} \in \overline{S} \mid \overline{\alpha} \in \overline{I} \setminus \overline{I}_{o} \}.$$

By Theorem IV.24 the set of all basic expression types is the set of those well-formed expression types whose abstract index is a basic index. By Theorem IV.25 the set of all functor types is the set of all those well-formed expression types whose abstract index is not basic.

Theorem IV.24 follows from Conclusion IV.21, Definition

II.6, Conclusion IV.19, Theorem IV.16a, Conclusion II.6a, Axiom II.2, Conventions II.3c and IV.3c and Conclusion IV.12. Theorem IV.25 follows from Conclusion IV.22, Definition II.7, Conclusion IV.19, Theorem IV.16a, Conclusion II.6a, Theorem I.5[d], Axiom II.2, Conventions II.3c and IV.3c, and Conclusions IV.13 and IV.12.

The concepts of the family $\overline{Ct}(\overline{S})$ of all syntactic categories of well-formed expression types of $\overline{\mathscr{L}}$, the family $\overline{Ct}(\overline{B})$ of all basic categories of $\overline{\mathscr{L}}$ and the family $\overline{Ct}(\overline{F})$ of all functor categories of $\overline{\mathscr{L}}$ are induced, respectively, by the concepts of $Ct(S)$, $Ct(B)$, and $Ct(F)$. Without writing out the definitions of those induced concepts we merely note that

REMARK IV.2. In **DTSCL** the definitions of the concepts $\overline{Ct}(\overline{S})$, $\overline{Ct}(\overline{B})$ and $\overline{Ct}(\overline{F})$ are expressions induced by Definitions II.11a,b,c, respectively.

We now join to the theory **DTSCL** two definitions of four-argument relations: the *relation* $(\bar{/})^n$ *of replacing a constituent of n-th order* (for $n \geq 0$) *of an expression type*, and the *relation* $(\bar{/})$ *of replacing a constituent of an expression type*.

DEFINITION IV.26a

$$\bar{r}(\bar{p}\bar{/}\bar{q})^{\bar{n}}\bar{s} \Leftrightarrow \bigvee_{p,q,r,s} (\bar{p}=[p] \wedge \bar{q}=[q] \wedge \bar{r}=[r] \wedge \bar{s}=[s] \wedge r(p/q)^n s).$$

DEFINITION IV.26

$$\bar{r}(\bar{p}\bar{/}\bar{q})\bar{s} \Leftrightarrow \bigvee_{p,q,r,s} (\bar{p}=[p] \wedge \bar{q}=[q] \wedge \bar{r}=[r] \wedge \bar{s}=[s] \wedge r(p/q)s).$$

By referring to the definitions adopted above we can prove

Theorem IV.26

a. $\quad \bar{r}(\overline{p}/\overline{q})^{0}\bar{s} \Leftrightarrow \bar{s},\bar{r} \in \bar{S} \wedge \bar{q} = \bar{s} \wedge \bar{p} = \bar{r},$

b. $\quad \bar{r}(\overline{p}/\overline{q})^{1}\bar{s} \Leftrightarrow \bar{s},\bar{r} \in \bar{S} \wedge \bigvee_{n\geq 1} \bigvee_{\substack{\bar{s}_0,\bar{s}_1,\ldots,\bar{s}_n \in \bar{S} \\ \bar{r}_0,\bar{r}_1,\ldots,\bar{r}_n \in \bar{S}}}$

$$(\bar{s} = \bar{p}(\bar{s}_0,\bar{s}_1,\ldots,\bar{s}_n) \wedge \bar{r} = \bar{p}(\bar{r}_0,\bar{r}_1,\ldots,\bar{r}_n) \wedge$$

$$\wedge \bigvee_{0\leq j\leq n} (\bar{q} = \bar{s}_j \wedge \bar{p} = \bar{r}_j \wedge \bigwedge_{\substack{k\neq j \\ 0\leq k\leq n}} \bar{s}_k = \bar{r}_k)),$$

c. $\quad k > 0 \Rightarrow (\bar{r}(\overline{p}/\overline{q})^{k+1}\bar{s} \Leftrightarrow \bigvee_{\bar{t},\bar{v}} (\bar{r}(\overline{v}/\overline{t})^{k}\bar{s} \wedge \bar{v}(\overline{p}/\overline{q})^{1}\bar{t})),$

d. $\quad \bar{r}(\overline{p}/\overline{q})\bar{s} \Leftrightarrow \bigvee_{n} \bar{r}(\overline{p}/\overline{q})^{n}\bar{s}.$ [31]

Theorems IV.26a-d are expressions induced, respectively, by Definitions II.14a-c and II.14d.

Theorem IV.26a follows immediately from Definitions IV.26a, II.14a and Conclusions IV.19 and IV.18.

Theorem IV.26d follows immediately from Definitions IV.26, II.14d and IV.26.

The proofs of Theorems IV.26b,c are given in the Annex.

From the analyses carried out above we obtain

Theorem IV.12*. *Every expression induced by an axiom or a definition of* TSCL *is a thesis in* DTSCL.

Theorem IV.12* can be proved by reference to Theorems IV.8* and IV.11* and IV.22a-c, Remark IV.1, Theorems IV.23-25, Remark IV.2 and Theorems IV.26a-d.

Theorem IV.12* yields

THEOREM IV.13*. *Every expression induced by a thesis of* **TSCL** *is a thesis in* **DTSCL**.

By availing ourselves of Theorem IV.13* we can state that in **DTSCL** we have a theorem which says that the family $\overline{Ct}(\overline{S})$ of all syntactic categories of well-formed expression types is a logical partition of the set $\overline{S}$ of all well-formed expression types of the syntactically categorial language $\overline{\mathscr{L}}$, that partition being determined by the relation $\underset{c}{\equiv}$ (cf. Theorem II.6a). The analoga of both **fttsc** (see Sec.II.1 and Theorems II.7a,b), that is the *fundamental theorems of the theory of syntactic categories for expression types of* $\overline{\mathscr{L}}$, are also valid in **DTSCL**.

Sec. IV.6. The dual theory DTSCω-L

The dual theory **DTSCω-L** makes it possible to analyse the ω-language $\overline{\mathscr{L}}$ of expression types, described by the system (ω-$\overline{\mathscr{L}}_c$) given in Sec.IV.1. Now **DTSCω-L** is obtained from **TSCω-L** by the joining to the latter of the definitions of all concepts induced by its primitive and derived concepts. Since **TSCω-L** is based on **TETk**, and the concepts induced by the primitive and derived concepts of the latter have already been defined in **TETp**, it is self-evident that the following metatheorems hold:

THEOREM IV.14*. *Every thesis (accepted sentence) of* **TSCω-L**, *and in particular of* **TETk**, *is a thesis in* **DTSCω-L**.

THEOREM IV.15*. *Every thesis (accepted sentence) of* **TETp** *is a thesis in* **DTSCω-L**.

These two metatheorems will hereafter be used tacitly.

It can also be seen that **DTSCω-L** can be obtained by joining to **TETp** the definitions of all concepts induced by the concepts introduced in Chapter III.

The first two definitions to be given here are definitions of the concepts induced by the concepts of the sets O and Vr, respectively.

DEFINITION IV.27

$$\overline{O} = O/_{\approx} .$$

DEFINITION IV.28

$$\overline{Vr} = Vr/_{\approx} .$$

The concepts thus defined are termed the *set of all operator types* and the *set of all variable types*, respectively.

Definition IV.27 is equivalent to

CONCLUSION IV.23

$$\overline{p} \in \overline{O} \Leftrightarrow \bigvee_{p \in O} \overline{p} = [p].$$

Definition IV.28 is equivalent to

CONCLUSION IV.24

$$\overline{p} \in \overline{Vr} \Leftrightarrow \bigvee_{p \in Vr} \overline{p} = [p].$$

From Conclusions IV.23 and IV.24, Axiom III.1 and Conclusion IV.16 we obtain immediately

THEOREM IV.27

$$\overline{O} \cup \overline{Vr} \subseteq \overline{E}_s.$$

The set of all operator types and the set of all variable types are subsets of the set of all simple expression types.

Theorem IV.27 is induced by Axiom III.1 of **TSCω-L**.

It is self-evident that

THEOREM IV.16a*. *The expressions induced by Axioms III.2 and III.3 of* **TSCω-L** *are theorems in* **DTSCω-L**.

The concept of an operator expression token has in **DTSCω-L** its analogon in the concept of an *operator expression type*. An operator expression type of the ω-language $\overline{\mathscr{L}}$ is defined as an element of the set $\overline{\rho}(\overline{O}\times\overline{Vr}\times\overline{E})$, that is as a compound expression type consisting, successively, from an operator type, a variable type, and an expression type. As in Chapter III, we adopt a metalinguistic definition which makes it possible to record operator expressions of the ω-language $\overline{\mathscr{L}}$ in a simpler form.

DEFINITION IV.1*

$$\overline{p} = \overline{\rho}^{\omega}(\overline{p}_0,\overline{p}_1,\overline{p}_2) \underset{\mathrm{df}}{=\!=} \overline{p} = \overline{\rho}(\overline{p}_0,\overline{p}_1,\overline{p}_2) \wedge \overline{p}_0\in\overline{O} \wedge \overline{p}_1\in\overline{Vr} \wedge \overline{p}_2\in\overline{E}.$$

The expression on the left side of this definition is read: $\overline{p}$ is an operator expression type consisting of the main operator $\overline{p}_0$, the indexical variable $\overline{p}_1$ of that operator, and the scope $\overline{p}_2$ of that operator.

Definition IV.1*, Conclusions IV.23, IV.24, IV.18, III.1a,b, II.5b, Theorem IV.17a and Definition III.1* yield

THEOREM IV.28

$$\overline{p} = \overline{\rho}^{\omega}(\overline{p}_0,\overline{p}_1,\overline{p}_2) \Leftrightarrow \bigvee_{p_0,p_1,p_2,p} (\overline{p}_0 = [p_0] \wedge \overline{p}_1 = [p_1] \wedge \wedge \overline{p}_2 = [p_2] \wedge \overline{p} = [p] \wedge p = \rho^{\omega}(p_0,p_1,p_2)).$$

An operator expression type consisting of a main operator $\overline{p}_0$, *its indexical variable* $\overline{p}_1$ *and its scope* $\overline{p}_2$ *is a class of abstraction of the relation* $\approx$, *determined by an operator expression token consisting of the following expression tokens: the main operator* p_0, *its indexical variable* p_1 *and its scope* p_2, *which are, respectively, representatives of the classes of abstraction that determine the expression types* $\overline{p}_0, \overline{p}_1$ *and* $\overline{p}_2$.

We adopt the following definition of the *relation* $(\overline{fv})$ *of being a free variable in an expression type*:

DEFINITION IV.29

$$\overline{p}(\overline{fv})\overline{q} \Leftrightarrow \bigvee_{p,q} (\overline{p} = [p] \wedge \overline{q} = [q] \wedge p(fv)q).$$

This definition yields the following theorem induced by Definition III.1:

THEOREM IV.29

$$\overline{p}(\overline{fv})\overline{q} \Leftrightarrow \overline{p} \in \overline{Vr} \wedge \overline{p} \in \overline{C}_{\overline{q}} \wedge \neg \bigvee_{\overline{r},\overline{r}_0,\overline{r}_1,\overline{r}_2} (\overline{r} = \overline{\rho}^{\omega}(\overline{r}_0,\overline{r}_1,\overline{r}_2) \wedge \wedge \overline{r} \in \overline{C}_{\overline{q}} \wedge \overline{p} \in \overline{C}_{\overline{r}_2} \wedge \overline{p} = \overline{r}_1).$$

The proof of this theorem is given in the Annex.

The concepts induced by the concepts of sets: ${}^{n}S$, S, $S^{n\omega}$, S^{ω}, B, F, defined in **TSCω-L**, are now defined as the corresponding quotients of those sets by the relation $\approx$. We accordingly have

DEFINITION IV.30

$$ {}^{n}\overline{S} = {}^{n}S/_{\approx} $$

and

CONCLUSION IV.25

$$ \overline{p} \in {}^{n}\overline{S} \Leftrightarrow \bigvee_{p \in {}^{n}S} \overline{p} = [p]; $$

DEFINITION IV.31

$$ \overline{S} = S/_{\approx} $$

and

CONCLUSION IV.26

$$ \overline{p} \in \overline{S} \Leftrightarrow \bigvee_{p \in S} \overline{p} = [p]; $$

DEFINITION IV.32

$$ \overline{S}^{n\omega} = S^{n\omega}/_{\approx} $$

and

CONCLUSION IV.27

$$ \overline{p} \in \overline{S}^{n\omega} \Leftrightarrow \bigvee_{p \in S^{n\omega}} \overline{p} = [p]; $$

DEFINITION IV.33

$$ \overline{S}^{\omega} = S^{\omega}/_{\approx} $$

and

CONCLUSION IV.28

$$ \overline{p} \in \overline{S}^{\omega} \Leftrightarrow \bigvee_{p \in S^{\omega}} \overline{p} = [p]; $$

DEFINITION IV.34

$$ \overline{B} = B/_{\approx} $$

and

CONCLUSION IV.29

$$\bar{\rho} \in \bar{B} \Leftrightarrow \bigvee_{\rho \in B} \bar{\rho} = [\rho];$$

DEFINITION IV.35

$$\bar{F} = F/_{\approx}$$

and

CONCLUSION IV.30

$$\bar{\rho} \in \bar{F} \Leftrightarrow \bigvee_{\rho \in F} \bar{\rho} = [\rho].$$

By referring to Conclusions IV.25 and IV.26 we prove the following theorems which are induced by Definitions III.2a,b and III.2c:

THEOREM IV.30

a. $${}^{0}\bar{S} = \bar{E}_{s},$$

b. $$\bar{p} \in {}^{k+1}\bar{S} \Leftrightarrow \bar{p} \in {}^{k}\bar{S} \vee \bigvee_{n \geq 1} \bigvee_{\bar{p}_0, \bar{p}_1, \ldots, \bar{p}_n \in {}^{k}\bar{S}}$$

$$(\bar{p} = \bar{\rho}(\bar{p}_0, \bar{p}_1, \ldots, \bar{p}_n) \wedge \bar{p}_0 \notin \bar{O} \wedge \bar{c}^{n+1}(\bar{\alpha}, \bar{\alpha}_1, \ldots, \bar{\alpha}_n, \bar{\alpha}_0)) \vee$$

$$\vee \bigvee_{\bar{p}_0, \bar{p}_1, \bar{p}_2, \bar{p}_3} (\bar{p} = \bar{\rho}^{\omega}(\bar{p}_0, \bar{p}_1, \bar{p}_2) \wedge \bar{p}_2 \in {}^{k}\bar{S} \wedge \bar{p}_3 (\overline{fv}) \bar{p}_2 \wedge$$

$$\wedge \bar{p}_3 = \bar{p}_1 \wedge \bar{c}^{3}(\bar{\alpha}, \bar{\alpha}_1, \bar{\alpha}_2, \bar{\alpha}_0)).$$

c. $$\bar{S} = \bigcup_{n=0}^{\infty} {}^{n}\bar{S}.$$

Theorem IV.30a follows immediately from Conclusion IV.25, Definition III.2a and Conclusion IV.16. Theorem IV.30c follows simply from Conclusion IV.26, Definition III.2c and Conclusion

IV.25. The proof of Theorem IV.30b is given in the Annex.

By availing ourselves of Conclusion IV.27 and Definition III.3 and by following the proof of Theorem IV.30b we easily find that the following theorem, induced by Definition III.3, is valid:

THEOREM IV.31

$$\bar{p} \in \bar{S}^{n\omega} \Leftrightarrow \bigvee_{k} \bigvee_{n\geq 1} \bigvee_{\bar{p}_0,\bar{p}_1,\ldots,\bar{p}_n \in {}^k\bar{S}} (\bar{p} = \bar{\rho}(\bar{p}_0,\bar{p}_1,\ldots,\bar{p}_n) \wedge$$

$$\wedge\ \bar{p}_0 \not\in \bar{O} \wedge \bar{c}^{n+1}(\bar{\alpha},\bar{\alpha}_1,\ldots,\bar{\alpha}_n,\bar{\alpha}_0)).$$

Likewise, by applying to Conclusion IV.28 and Definition III.4 and by following the proof of Theorem IV.30b we find that the following theorem, induced by Definition III.4 is valid:

THEOREM IV.32

$$\bar{p} \in \bar{S}^{\omega} \Leftrightarrow \bigvee_{k} \bigvee_{\bar{p}_0,\bar{p}_1,\bar{p}_2,\bar{p}_3} (\bar{p} = \bar{\rho}^{\omega}(\bar{p}_0,\bar{p}_1,\bar{p}_2) \wedge \bar{p}_2 \in {}^k\bar{S} \wedge$$

$$\wedge\ \bar{p}_3 (\bar{f}\bar{v}) \bar{p}_2 \wedge \bar{p}_3 = \bar{p}_1 \wedge \bar{c}^3(\bar{\alpha},\bar{\alpha}_1,\bar{\alpha}_2,\bar{\alpha}_0)).$$

From Conclusion IV.29, Definition III.5, Conclusions IV.26 and IV.12, Theorem IV.16a, Conclusion III.6 and II.2b, Conventions II.3c and IV.3c and Axiom II.2 we obtain

THEOREM IV.33

$$\bar{B} = \{ \bar{p} \in \bar{S} \mid \bar{\alpha} \in \bar{I}_o \}.$$

From Conclusion IV.30, Definition III.6, Conclusions IV.26, IV.13 and IV.12, Theorem IV.16a, Conclusions III.6, Conventions II.3c and IV.3c and Axiom II.2 we obtain

THEOREM IV.34

$$\overline{F} = \{ \overline{p} \in \overline{S} \mid \overline{\alpha} \in \overline{I} \setminus \overline{I}_o \}.$$

The last two theorems are induced by Definitions III.5 and III.6, respectively.

The following two expressions, induced by the last two axioms of **TSCω-L**, respectively, are theorems in **DTSCω-L**:

THEOREM IV.35

$$\overline{\iota}(\overline{S}^{n\omega}) \cap \overline{I}_o \neq \emptyset$$

and

THEOREM IV.36

$$\overline{\iota}(\overline{S}^{\omega}) \cap \overline{I}_o \neq \emptyset.$$

Theorem IV.35 follows from Axiom III.4, Conclusions IV.27, IV.12, III.5a and II.5a and Theorem IV.16a. Theorem IV.36 follows immediately from Axiom III.5, Conclusions IV.28, IV.12, III.5b and II.5a and Theorem IV.16a.

In **TSCω-L** we have adopted definitions of the families $Ct(S)$, $Ct(B)$, $Ct(F)$ and definitions of the relations $(/)^n$ and $(/)$. We now assume that

REMARK IV.3. In **DTSCω-L** the concepts $\overline{Ct}(\overline{S})$, $\overline{Ct}(\overline{B})$ and $\overline{Ct}(\overline{F})$ are defined by expressions induced by Definitions III.7a,b,c, respectively.

The expressions $\overline{r}(\overline{p}/\overline{q})^n\overline{s}$ and $\overline{r}(\overline{p}/\overline{q})\overline{s}$ are defined thus:

DEFINITION IV.36a

$$\overline{r}(\overline{p}/\overline{q})^n\overline{s} \Leftrightarrow \bigvee_{p,q,r,s} (\overline{p} = [p] \wedge \overline{q} = [q] \wedge \overline{r} = [r] \wedge \overline{s} = [s] \wedge r(p/q)^n s).$$

DEFINITION IV.36

$$\overline{r}(\overline{p}/\overline{q})\overline{s} \Leftrightarrow \bigvee_{p,q,r,s} (\overline{p}=[p] \wedge \overline{q}=[q] \wedge \overline{r}=[r] \wedge \overline{s}=[s] \wedge r(p/q)s\ ).$$

Definitions IV.36a, III.8a, Conclusion IV.26 and Axiom I.1a immediately yield the theorem which is induced by Definition III.8a. The theorem induced by Definition III.8b also holds in **DTSCω-L**. Its proof, similar to the proof of Theorem IV.26b, is based on Definitions IV.36a and III.8b, Theorems IV.17a and IV.28, Conventions II.3a and IV.3a, Conclusions IV.26 and IV.18 and Theorems III.3a,b. The proof of the theorem induced by Definition III.8c is modelled on the proof of Theorem IV.26c. The theorem induced by Definition III.8d follows immediately from Definitions IV.36, III.8d and IV.36a. We can accordingly state that

THEOREM IV.16*b. *Expressions induced by Definitions III.8a–c and III.8d are theorems in* **DTSCω-L**.

In view of what has been said above and referring to Theorems IV.8*, IV.15*, IV.27, IV.16*a, IV.29–IV.36, Remark IV.3 and Theorem IV.16*b we can state that

THEOREM IV.16*. *Every expression induced by an axiom or a definition of* **TSCω-L** *is a thesis in* **DTSCω-L**.

This leads us to

THEOREM IV.17*. *Every expression induced by a thesis of* **TSCω-L** *is a thesis in* **DTSCω-L**.

By availing ourselves of the above metatheorem we can state

in particular that in the dual theory **DTSCω-L** we have the theorem stating that the family $\overline{Ct}(\overline{S})$ of all syntactic categories of well-formed expression types of the ω-language $\overline{\mathscr{L}}$ is a logical classification of the set $\overline{S}$ of all well-formed expression types of that language, determined by the relation $\underset{c}{\equiv}$ of categorial agreement (cf. Theorem III.5). The said metatheorem also authorizes us to state that the theorems induced by both **fttsc** given in Chap.III (see Theorems III.6a,b), that is *fundamental theorems of the theory of syntactic categories for expression types of the ω-language* $\overline{\mathscr{L}}$ hold in **DTSCω-L**. The ω-language $\overline{\mathscr{L}}$ can, therefore, correctly be referred to as a syntactically categorial language.

FINAL REMARKS

We have presented two formal axiomatic conceptions of languages pertaining to simple languages and ω-languages, respectively, i.e., languages which contain operators and variables bound by them. Both theories have been based on the theory of syntactic categories of Leśniewski-Ajdukiewicz-Curry. In working out of both conceptions we have taken into consideration that trend in the ontology of language which refers to *nominalism*. We have done so because, without calling in question the fact that in logical and semiotic practice we accord to linguistic objects a dual ontological status, that is to say, recognizing the fact that

1° on the one hand, inscriptions, including words and expressions, are physical objects that have a definite shape, are extensive in space and time and accessible to sensory cognition, in a word, are *concrete* entities,

2° on the other, linguistic entities, including words and expressions, are classes of equiform material objects, and are thus abstract, *ideal*, entities,

in formalizing both theories we started from the assumption that *tokens*, and hence concrete objects, are the basic stratum of language and we postulated their existence as primitive entities. *Types*, that is abstract objects of a different ontological nature, were assigned derived existence and defined by reference to tokens.

Thus, taking into consideration the *nominalistic* trend,

which is otherwise in agreement with the views of Stanisław Leśniewski and ontological reism of Tadeusz Kotarbiński, we formalized both theories at *two different levels*, that of *tokens* and that of *types*, starting from the *token level*. We accordingly constructed first axiomatic theories of expression tokens (*token level*), and next (as their definitional extension) at the *type level* we formulated dual theories of languages of expression types.

We demonstrated that the passing from the theories of languages of expression tokens to the corresponding dual theories of languages of expression types is quite natural. The important point is that the concepts at the *type level*, i.e., set of types, object types, and relations among them, that is concepts induced relative to the corresponding concepts at the *token level*, i.e., sets of tokens, object tokens, and relations among them, inherit all the properties of the latter. This is so because every axiom, every definition, and hence also every theorem of a given theory of the language of expression tokens, which describes property or properties at the *token level*, when translated into an induced expression which describes the property or properties of the corresponding induced concept(s) at the *type level*, is a theorem of the dual theory, hence a theory which also describes language categorially, but as a language of expression types.

In the light of the above it is self-evident that both theories of the syntax of languages of expression types can be based on axioms and definitions which are expressions induced by the axioms and definitions of the corresponding theories of languages of expression tokens. Hence in purely theoretical considerations pertaining to the syntactic and categorial description of language we can disregard the ontological nature of its objects. In such a case the postulation at the outset of the existence of linguistic concrete objects (tokens) is not necessary. But from the point of view of philosophy the adoption of such an assumption and the basing of the theory of the syntax of language on it is of essential importance because *it points to the possibility of avoiding the postulation of the existence of abstract linguistic objects in syntactic analyses*.

The above philosophical thesis, which is a result of considerations presented in this book, can be substantiated much more strongly, which is shown by the latest studies carried out by the present author.

We shall brieffy substantiate the opinion voiced above. As it is known, the logical conception of the ontology of the syntax of language should provide formal foundations that would enable one to formalize the theory of language at two levels by taking into account the dual nature of linguistic objects. The theory of language can thus be constructed if we start from the assumption that the study of language is based on object types, that is ideal entities intuitively interpreted as having representations in tokens, that is concretes treated as entities derived from the former. In such a case we obviously refer to a different philosophical trend, namely Platonism, and we begin the formalization of the theory of the syntax of language at the *type level* by postulating the existence of object types and constructing first the theory of expression types. Later, by making use of its extension, at the second level (*token level*) we construct the theory of the syntax of language of expression tokens by defining the concepts at that level by reference to types.

Then the problem arises: would the descriptive potential of such a theory of the syntax of language be an argument to the detriment of that theory of the syntax of language in the formalization of which the *nominalistic* approach was used?

Let the nature of that problem be explained by an example. The relation of concatenation between types is a function, whereas when defined on tokens it is not a function. Should not that fact be of an essential importance in the description of the syntax of language based on the *Platonizing* approach? Does it not allow one to describe some essentially new, and even different, properties of objects at the *type level*, and new or different properties of objects at the *token level*?

The answer to this essential question is in the negative if we take into account the categorial approach to language, discussed in this book. This is demonstrated by the studies carried out by the present author, indicated in [1986 and 1990], and developed in [1988 and 1989]. The example of two formal theories of simple categorial languages, which represent the outlined opposing *nominalistic* and *Platonizing* approach to the syntax of language, respectively, demonstrates that the set of theses (axioms, definitions, and theorems) of both theories is the same: the said *theories are equivalent*. Moreover, it is shown that every property of concepts at the *token* (or respectively *type*) *level*, described by a thesis of one theory, has its analogue in the corresponding property of concepts at the *type* (or respectively *token*) *level*, described by a thesis of the other theory. And conversely, every property of the corresponding concepts at the *type* (or respectively *token*) *level* described by a thesis of the other theory (i) either has its analogue in a property of the corresponding concepts at the *token* (or respectively *type*) *level*, described by the thesis of the first theory, (ii) or can be transposed into such a property which is described by a thesis of the first theory. Thus those theories are in a sense, *mutually interpretable* and allow for a two-level framework of the syntax of simple languages in an analogical manner, that is categorially, even though the theories under consideration start from opposing ontological assumptions in the matter of the nature of linguistic objects.

The equivalence of both theories of the syntax of simple languages and their mutual interpretability thus consolidate the philosophical thesis stating that in syntactic research on simple languages approached categorially one can eliminate the assumption of the existence of ideal linguistic objects. This conclusion can be formulated also with reference to the theory of ω-languages described categorially, and probably also with reference to other theories of language, for instance the theories of formal languages in Chomsky's spirit.

ANNEX

We give here the proofs of certain theorems formulated in the main text. The numbering of the theorems and their order in the text are observed.

THEOREM I.2

$$c(p,q,t) \wedge c(r,s,u) \Rightarrow (t \approx u \Leftrightarrow p \approx r \wedge q \approx s \vee$$
$$\vee \bigvee_v (c(r,v,p) \wedge c(v,q,s) \vee c(p,v,r) \wedge c(v,s,q))).$$

Now Theorem I.2 follows from two theorems:

THEOREM I.2a

$$c(p,q,t) \wedge c(r,s,u) \wedge t \approx u \Rightarrow p \approx r \wedge q \approx s \vee$$
$$\vee \bigvee_v (c(r,v,p) \wedge c(v,q,s) \vee c(p,v,r) \wedge c(v,s,q))$$

and

THEOREM I.2b

$$c(p,q,t) \wedge c(r,s,u) \wedge (p \approx r \wedge q \approx s \vee$$
$$\vee \bigvee_v (c(r,v,p) \wedge c(v,q,s) \vee c(p,v,r) \wedge c(v,s,q))) \Rightarrow t \approx u.$$

P r o o f of Theorem I.2a. Let

(1) $$c(p,q,t) \wedge c(r,s,u) \wedge t \approx u.$$

From Axiom I.8 and the formula (1) we obtain

$$p \approx r \vee \bigvee_v (c(r,v,p) \vee c(p,v,r)).$$

If p is equiform with r, then by assumption (1) and Conclusion I.2 q is equiform with s and the thesis of the theorem is true. Let us, therefore, assume additionally that

$c(r,v_1,p) \vee c(p,v_1,r)$ and consider the case when

(1.1) $$c(r,v_1,p).$$

It follows from (1) and (1.1) that

(1.2) $$c(r,v_1,p) \wedge c(p,q,t).$$

Further, by Axiom I.2, we have

(1.3) $$c(v_1,q,v_2) \wedge c(r,v_2,t_1).$$

Hence, by Axiom I.5, formulas (1.2) and (1.3) yield $t \approx t_1$ so that, by (1.3) and Axiom I.7, it follows that

(1.4) $$c(r,v_2,t),$$

and since $c(r,s,u) \wedge t \approx u$ by assumption (1), by Axiom I.7 we obtain

(1.5) $$c(r,s,t).$$

In turn (1.4) and (1.5) and Axioms I.1a and I.6 yield

(1.6) $$s \approx v_2.$$

By applying Axiom I.7 once more to (1.3) and (1.6) we obtain the relationship $c(v_1,q,s)$, which together with (1.1) yields

(1.7) $$c(r,v_1,p) \wedge c(v_1,q,s).$$

The thesis of the theorem under consideration follows immediately from (1.7). We have still to analyse the case when $c(p,v_1,r)$. Since in that case the proof of the thesis is analogical as in the case when $c(r,v_1,p)$, it is omitted here.

We now proceed to prove Theorem I.2b. Let us assume that

(1) $$c(p,q,t) \wedge c(r,s,u)$$

and

(2) $$p \approx r \wedge q \approx s \vee \bigvee_{v} (c(r,v,p) \wedge c(v,q,s) \vee$$
$$\vee\ c(p,v,r) \wedge c(v,s,q)).$$

If the first element of the disjunction (2) holds the thesis of the theorem follows immediately from (1) and Axiom I.4. Let us, therefore, assume that

$$c(r,v_2,p) \wedge c(v_2,q,s) \vee c(p,v_2,r) \wedge c(v_2,s,q)$$

and consider the case when

(1.1) $$c(r,v_2,p) \wedge c(v_2,q,s).$$

By assumptions (1) and (1.1) we obtain

(1.2) $$c(r,v_2,p) \wedge c(p,q,t) \wedge c(v_2,q,s) \wedge c(r,s,u).$$

By availing ourselves of Axiom I.5 we obtain immediately from (1.2) that $t \approx u$, which concludes the proof for the case (1.1). In the other case, when $c(p,v_2,r) \wedge c(v_2,s,q)$, the proof of the thesis is analogical. It is therefore omitted here.

THEOREM I.7

$$^{n}V \subseteq W.$$

The proof of this theorem is by induction with respect to the variable n.

P r o o f. For n=1 the theorem is true in view of Definition I.2a and Conclusion I.3a. Let us, therefore, assume inductively that $^{k}V \subseteq W$ and, additionally, that $r \in {}^{k+1}V$. Then, by Definition I.2b, $p_1 \in {}^{k}V$, $q_1 \in V$, and $c(p_1,q_1,r)$. By the inductive assumption $p_1 \in W$, and by Conclusion I.3a, $q_1 \in W$. By availing ourselves of Conclusion I.3b we have $r \in W$. Thus it follows from the assumption ${}^{k}V \subseteq W$ that ${}^{k+1}V \subseteq W$, and since the theorem is true for $n=1$ it is true for any natural number n.

THEOREM I.8

$$p \in {}^{n}V \wedge q \approx p \Rightarrow q \in {}^{n}V.$$

P r o o f. For $n=1$ the theorem follows immediately from Axiom I.10 and Definition I.2a. Assume that $n>1$, $p \in {}^{n}V$ and $q \approx p$. It follows from Definition I.2b that $c(p_1,p_2,p)$, $p_1 \in {}^{n-1}V$ and $p_2 \in V$. Since $q \approx p$, by Axiom I.7 we obtain $c(p_1,p_2,q)$. Hence, by Definition I.2b, $q \in {}^{n}V$, which was to be demonstrated.

THEOREM I.9

$$p \in {}^{m}V \wedge q \in {}^{n}V \wedge c(p,q,r) \Rightarrow r \in {}^{m+n}V.$$

The inductive proof of this theorem is carried out with respect to the variable n.

P r o o f. Note first that for $n=1$ the theorem is true in view of Definition I.2a,b. Let us, therefore, assume inductively that it is true for $n=k$ and demonstrate its validity for $n=k+1$. We accordingly assume in addition that

(1) $$p \in {}^{m}V,$$

(2) $$q \in {}^{k+1}V,$$

(3) $$c(p,q,r).$$

By Definition I.2b we obtain from (2)

(4) $$q_1 \in {}^{k}V,$$

(5) $$q_2 \in V,$$

(6) $$c(q_1,q_2,q).$$

By availing ourselves of Axiom I.2 we obtain

(7) $$c(p,q_1,r_1),$$

(8) $$c(r_1,q_2,r_2).$$

It follows from the inductive assumption and formulas (1), (4) and (7) that $r_1 \in {}^{m+k}V$. Since the theorem under consideration is true for $n=1$, the above conclusion, the formula (5), Definition I.2a and formula (8) yield

(9) $$r_2 \in {}^{m+k+1}V.$$

By referring to Theorem I.2 and to formulas (8) and (3), and (7) and (6), in that order, we find that

(10) $$r_2 \approx r.$$

Hence (10), Axiom I.1b and (9), in view of Theorem I.8, yield $r \in {}^{m+k+1}V$, which concludes the proof.

THEOREM I.10

$$r \in {}^{n+1}V \Leftrightarrow \bigvee_{p \in V} \bigvee_{q \in {}^{n}V} c(p,q,r).$$

P r o o f. We shall first prove the lemma in which the above equivalence is replaced by the implication

$$r \in {}^{n+1}V \Rightarrow \bigvee_{p \in V} \bigvee_{q \in {}^{n}V} c(p,q,r).$$

The proof of this lemma is by induction with respect to the variable n.

For $n=1$ the lemma is true by Definition I.2a,b. Let us assume the truth of the lemma for $n=k$. We shall prove that it holds for $n=k+1$. Assume accordingly that $r \in {}^{k+1+1}V$. By Definitions I.2b,a we have

(1) $$p_1 \in {}^{k+1}V,$$

(2) $$p_2 \in {}^{1}V,$$

(3) $$c(p_1,p_2,r),$$

and in view of the inductive assumption and (1) we have

(4) $$q_1 \in V,$$

(5) $$q_2 \in {}^{k}V,$$

(6) $$c(q_1,q_2,p_1).$$

We now apply Axiom I.2 to the labels q_2 and p_2 and again to the labels q_1 and the concatenation of q_2 and p_2. We obtain

(7) $$c(q_2,p_2,t_1),$$

(8) $$c(q_1,t_1,r_1).$$

Now $r \approx r_1$ follows from Axiom I.5 and formulas (6), (3), (7) and (8). Hence $c(q_1,t_1,r)$ in view of (8) and Axiom I.7. Since $t_1 \in {}^{k+1}V$ follows from Theorem I.9 and (5), (2) and (7), in view of (4) we can state that

$$\bigvee_{p \in V} \bigvee_{q \in {}^{k+1}V} c(p,q,r).$$

Thus on the assumption that the lemma is true for $n=k$ we have shown its truth for $n=k+1$, which concludes its inductive proof.

It can easily be seen that the implication which is the converse of the lemma proved above is also true, so that Theorem I.10 is true by Theorem I.9 and Definition I.2a.

THEOREM I.11b

$$\rho \in W \Rightarrow \bigvee_{n}^{1} \rho \in {}^{n}V.$$

P r o o f. By Theorem I.11a, $\rho \in W \Rightarrow \bigvee_{n} \rho \in {}^{n}V$. Hence it suffices to demonstrate that $\rho \in {}^{m}V \wedge \rho \in {}^{n}V \Rightarrow m=n$. The proof of this lemma will be by induction with respect to the variable n.

The lemma is true for $n=1$, for if it could be so that $\rho \in {}^{m}V$, $\rho \in {}^{1}V$ and $m \neq 1$, then by Definitions I.2a,b, Convention I.3 and Axiom I.11 ρ, on the one hand, would be a concatenation of two word tokens (because $\rho \in {}^{m}V$), and, on the other, would not be a concatenation of any pair of labels (because $\rho \in {}^{1}V$), which results in a contradiction. Let us assume inductively that the lemma is true for $n=k$. We shall prove its truth for $n=k+1$. Let accordingly

(1) $$\rho \in {}^{m}V,$$

(2) $$\rho \in {}^{k+1}V.$$

By Definition I.2b, we obtain - in view of (2) - the expressions

(3) $$\rho_1 \in {}^{k}V,$$

(4) $$\rho_2 \in V,$$

(5) $$c(\rho_1, \rho_2, \rho).$$

Now (5) yields $\rho \notin V$ by Axiom I.11, and hence also $\rho \notin {}^{1}V$ in view of Definition I.2a. Since (1) holds, $m>1$. We can, therefore, apply Definition I.2b to formula (1). Then we obtain the relationships

(6) $$t_1 \in {}^{m-1}V,$$

(7) $$t_2 \in V,$$

(8) $$c(t_1,t_2,p).$$

We now apply Theorem I.3 to (5), (8), (4) and (7) and obtain

(9) $$p_1 \approx t_1,$$

while by Theorem I.8 from (6) and (9) we obtain

(10) $$p_1 \in {}^{m-1}V.$$

We now apply the inductive assumption to (10) and (3). Then we have $m-1=k$, and hence $m=k+1$, which was to be demonstrated.

THEOREM I.12

$$r\in W \wedge |r| = n > 1 \Leftrightarrow \bigwedge_{\substack{k,l \\ k+l=n}} \bigvee_{p,q\in W} (|p| = k \wedge |q| = l \wedge c(p,q,r)).$$

P r o o f. We shall carry out only an inductive proof of the simple implication, because the converse implication follows directly from Conclusions I.3b and I.7. Assume accordingly that $r\in W$ and $|r|=n=2$. Then by Definition I.3 $r\in{}^2V$ and in view of Definitions I.2b,a and I.3 and Theorem I.7 the consequent of the implication is true. Assume inductively that the implication under consideration is true for $n=m>1$. We shall demonstrate that it is true for $n=m+1$. Let accordingly

(1) $$r \in W \text{ and } |r| = m+1.$$

It follows from Definition I.3 and (1) that $r\in{}^{m+1}V$, which - in view of Definition I.2b - yields

(2) $$r_1 \in {}^{m}V,$$

(3) $$r_2 \in V,$$

(4) $$c(r_1, r_2, r).$$

Further, in view of Theorem I.10, we have

(5) $$t_1 \in V,$$

(6) $$t_2 \in {}^{m}V,$$

(7) $$c(t_1, t_2, r).$$

The inductive assumption yields, in view of (2), Theorem I.7 and Definition I.3,

(8) $$\bigwedge_{\substack{k,l \\ k+l=m}} \bigvee_{p,q\in W} (|p| = k \wedge |q| = l \wedge c(p,q,r_1)),$$

and in view of (6)

(9) $$\bigwedge_{\substack{k,l \\ k+l=m}} \bigvee_{p,q\in W} (|p| = k \wedge |q| = l \wedge c(p,q,t_2)).$$

Assume additionally that $i+j=m+1$. Then $(i-1)+j=m \wedge i-1\geq 1$ or $i+(j-1)=m \wedge j-i\geq 1$.

Let us consider first the case when the first part of the above disjunction holds so that

(1.1) $$(i-1)+j = m \wedge i-1 \geq 1.$$

In this case, in view of (9), we have

(1.2) $$p_1, q_1 \in W \wedge |p_1| = i-1 \wedge |q_1| = j$$

and

(1.3) $$c(p_1, q_1, t_2).$$

By applying Axiom I.2 we obtain

(1.4) $$c(t_1,p_1,u_1),$$

(1.5) $$c(u_1,q_1,u_2),$$

which yield $r \approx u_2$ in view of (1.3) and (7) and Axiom I.5. This and (1.5) yield, by Axiom I.7,

(1.6) $$c(u_1,q_1,r).$$

Since $|u_1|=i$ and $u_1 \in W$ in view of (5), (1.2), (1.4), Conclusions I.3a,b and I.7 and Definitions I.2a and I.3, and $|q_1|=j$ and $q_1 \in W$ in view of (1.2), we can state, considering (1.6) that

(1.7) $$\bigvee_{p,q \in W} (|p| = i \wedge |q| = j \wedge c(p,q,r)).$$

We consider now the second case, that is

(2.1) $$i+(j-1) = m \wedge j-1 \geq 1.$$

By availing ourselves of formula (8) we obtain from the above assumption that

(2.2) $$p_2,q_2 \in W \wedge |p_2| = i \wedge |q_2| = j-1$$

and

(2.3) $$c(p_2,q_2,r_1).$$

By applying Axiom I.2 we have

(2.4) $$c(q_2,r_2,v_1),$$

(2.5) $$c(p_2,v_1,v_2).$$

By availing ourselves of Axiom I.5, we obtain $r \approx v_2$ in view of (2.3), (4), (2.4) and (2.5). Hence, in view of (2.5) and Axiom I.7, we have

(2.6) $$c(p_2,v_1,r).$$

It follows from (2.2) that $p_2 \in W$ and $|p_2|=i$, whereas $v_1 \in W$ and $|v_1|=j$ follows from Conclusions I.3a,b and I.7, formulas (2.2) and (3), Definitions I.2a and I.3 and formula (2.4). Hence in view of (2.6) we can state that

(2.7) $$\bigvee_{p,q \in W} (|p| = i \wedge |q| = j \wedge c(p,q,r)).$$

Thus for the assumption that $i+j=m+1$ in both cases under consideration, i.e., (1.1) and (2.1), we have

$$\bigvee_{p,q \in W} (|p| = i \wedge |q| = j \wedge c(p,q,r)),$$

and hence

$$\bigwedge_{\substack{i,j \\ i+j=m+1}} \bigvee_{p,q \in W} (|p| = i \wedge |q| = j \wedge c(p,q,r)).$$

The last expression proves that the assumption of the truth of the implication for $n=m>1$ results in its being true for $n=m+1$. This concludes the inductive proof of that implication.

THEOREM I.13a

$$p \in W \wedge |p| = n+1 \Leftrightarrow \bigvee_{p_1, p_2, \ldots, p_{n+1} \in V} c^{n+1}(p_1,p_2,\ldots,p_{n+1},p).$$

P r o o f. We refer to the principle of mathematical induction. For $n=1$ the expression $p \in W \wedge |p|=2$ is equivalent to $p \in {}^2V$ by Definition I.3 and Theorem I.7, and that in turn is equivalent, by Definitions I.2b,a, to the expression

$$\bigvee_{p_1, p_2 \in V} c(p_1,p_2,p).$$

Thus, in view of Definition I.4a, the theorem is true for $n=1$.

We assume that it is true for $n=k\geq 1$ and shall demonstrate its truth for $n=k+1$. For that purpose we write out a sequence of equivalences which follow from Definitions I.3 and I.2b, Theorem I.7, the inductive assumption and Definition I.4b:

$$p \in W \wedge |p| = k+2 > 1 \Leftrightarrow p \in {}^{k+2}V \Leftrightarrow \bigvee_{q\in {}^{k+1}V} \; \bigvee_{p_{k+2}\in V} c(q,p_{k+2},p) \Leftrightarrow$$

$$\Leftrightarrow \bigvee_{q} \; \bigvee_{p_1,p_2,\ldots,p_{k+2}\in V} (c^{k+1}(p_1,p_2,\ldots,p_{k+1},q) \wedge c(q,p_{k+2},p)) \Leftrightarrow$$

$$\Leftrightarrow \bigvee_{p_1,p_2,\ldots,p_{k+1},p_{k+2}\in V} (c^{k+2}(p_1,p_2,\ldots,p_{k+1},p_{k+2},p).$$

It follows from these equivalences that the inductive assumption implies that the theorem under consideration is true for $n=k+1$. Hence the theorem is true for any natural number n.

THEOREM I.14

$$c^{n+1}(p_1,p_2,\ldots,p_{n+1},p) \wedge c^{n+1}(q_1,q_2,\ldots,q_{n+1},p) \wedge$$

$$\wedge \bigwedge_{1\leq i\leq n+1} p_i,q_i \in V \Rightarrow \bigwedge_{1\leq i\leq n+1} p_i \approx q_i.$$

P r o o f. It follows immediately from Theorem I.3 and Definition I.4a that the theorem is true for $n=1$. We assume that it is true for $n=k$. To demonstrate that it is then true for $n=k+1$ we assume additionally that

(1) $\quad c^{k+2}(p_1,p_2,\ldots,p_{k+1},p_{k+2},p) \wedge c^{k+2}(q_1,q_2,\ldots,q_{k+1},q_{k+2},p)$

and

(2) $$\bigwedge_{1\leq i\leq k+2} p_i,q_i \in V.$$

We shall demonstrate that

$$\bigwedge_{1 \le i \le k+2} p_i \approx q_i.$$

In view of (1) it follows from Definition I.4b that

(3) $$c^{k+1}(p_1, p_2, \ldots, p_{k+1}, t_1) \wedge c^{k+1}(q_1, q_2, \ldots, q_{k+1}, t_2)$$

and

(4) $$c(t_1, p_{k+2}, p) \wedge c(t_2, q_{k+2}, p).$$

By Theorem I.3, formulas (4) and (2) yield

(5) $$t_1 \approx t_2$$

and

(6) $$p_{k+2} \approx q_{k+2}.$$

In view of the theorem which is a generalization of Axiom I.7 (see Theorem I.15) formulas (3), (5) and (2) yield

$$c^{k+1}(p_1, p_2, \ldots, p_{k+1}, t_1) \wedge c^{k+1}(q_1, q_2, \ldots, q_{k+1}, t_1) \wedge$$

$$\wedge \bigwedge_{1 \le i \le k+1} p_i, q_i \in V,$$

whence by the inductive assumption we obtain

$$\bigwedge_{1 \le i \le k+1} p_i \approx q_i.$$

Since, moreover, (6) holds, we note that

$$\bigwedge_{1 \le i \le k+2} p_i \approx q_i,$$

which concludes the inductive proof of the theorem under consideration.

THEOREM I.18

$$c^{n+1}(p_1,p_2,\ldots,p_{n+1},p) \wedge c^{n+1}(q_1,q_2,\ldots,q_{n+1},q) \wedge$$

$$\wedge\ p \approx q \wedge 1 \leq j \leq n+1 \wedge \bigwedge_{\substack{1\leq i\leq n+1\\ i\neq j}} (p_i \approx q_i) \Rightarrow p_j \approx q_j.$$

The proof is by induction with respect to the variable n.

P r o o f. Note first that for $n=1$ the theorem follows immediately from Conclusion I.2 and Definition I.4a. Let us, therefore, assume its truth for $n=k$ and demonstrate that it holds for $n=k+1$. For that purpose we assume additonally that

(1) $c^{k+2}(p_1,p_2,\ldots,p_{k+1},p_{k+2},p) \wedge c^{k+2}(q_1,q_2,\ldots,q_{k+1},q_{k+2},q)$

and

(2) $$p \approx q,$$

and also

(3) $$1 \leq j \leq k+2$$

and

(4) $$\bigwedge_{\substack{1\leq i\leq k+2\\ i\neq j}} p_i \approx q_i.$$

It follows from Definition I.4b and formula (1) that

(5) $$c^{k+1}(p_1,p_2,\ldots,p_{k+1},t_1) \wedge c^{k+1}(q_1,q_2,\ldots,q_{k+1},t_2)$$

and

(6) $$c(t_1,p_{k+2},p) \wedge c(t_2,q_{k+2},q).$$

Let us consider now two cases, namely when $j=k+2$ and when $1\leq j\leq k+1$.

In the first case we see, in view of (4) that

$$\bigwedge_{1\leq i\leq k+1} p_i \approx q_i.$$

By Theorem I.17, in view of (5), we have $t_1 \approx t_2$, and since (2) holds from Conclusion I.2 and formula (6) we obtain $p_{k+2} \approx q_{k+2}$. Thus

(7) $$j = k+2 \Rightarrow p_j \approx q_j.$$

In the second case ($1\leq j\leq k+1$), we have by (4)

$$p_{k+2} \approx q_{k+2} \text{ and } \bigwedge_{\substack{1\leq i\leq k+1 \\ i\neq j}} p_i \approx q_i.$$

Hence, by Conclusion I.2 and formulas (2) and (6), $t_1 \approx t_2$. Hence also, by the inductive assumption and formula (5), we have $p_j \approx q_j$ in view of

$$\bigwedge_{\substack{1\leq i\leq k+1 \\ i\neq j}} p_i \approx q_i.$$

Therefore

(8) $$1 \leq j \leq k+1 \Rightarrow p_j \approx q_j.$$

From the implications (7) and (8) and formula (3) we obtain self-evidently that $p_j \approx q_j$ and that the thesis of Theorem I.18 for $n=k+1$ is true. This concludes the inductive proof of Theorem I.18.

THEOREM II.1

$$W \cap I = \emptyset.$$

P r o o f. Let us assume that $p \in W \cap I$ so that $p \in W \wedge p \in I$. Then, in view of Axioms I.9 and II.1 and Conclusions I.3a, I.3^{d}a, I.4b and I.4^{d}b we have one of the following cases:

(i) $$p \in V \wedge p \in I_o,$$

(ii) $$p \in W \setminus V \wedge p \in I_o,$$

(iii) $$p \in V \wedge p \in I \setminus I_o,$$

(iv) $$p \in W \setminus V \wedge p \in I \setminus I_o.$$

Now case (i) is excluded by Axiom II.6. Cases (ii) and (iii) are excluded, respectively, by Axioms I.12 and II.3 and Axioms II.4 and II.11. We have, therefore, to analyse case (iv). It follows from Theorems I.11 and I.11^d - in view of the condition (iv), Definition I.2a and the definition dual to the latter - that $p \in {}^m V$ and $p \in {}^n I$, where $m>1$ and $n>1$. Hence it follows, by Theorems I.10 and I.10^d, that there are p_1, q_1, p_2, q_2 such that

$$p_1 \in V \wedge q_1 \in {}^{m-1}V \wedge c(p_1,q_1,p) \wedge p_2 \in I_o \wedge$$
$$\wedge\ q_2 \in {}^{n-1}I \wedge c(p_2,q_2,p).$$

By applying Axiom I.8 to the formula $c(p_1,q_1,p) \wedge c(p_2,q_2,p)$ we find - in view of the fact that $p_1 \in V$ and $p_2 \in I_o$ and in view of Axioms I.11 and II.3 - that $p_1 \approx p_2 \wedge p_1 \in V \wedge p_2 \in I_o$. Hence it follows by Axiom I.10 that $p_2 \in V$ and $p_2 \in I_o$, which is in contradiction with Axiom II.6. This concludes the indirect proof of Theorem II.1.

CONCLUSION II.3

$$p = \rho(p_0,p_1,\dots,p_n) \wedge q = \rho(q_0,q_1,\dots,q_n) \Rightarrow$$
$$\Rightarrow (p \approx q \Leftrightarrow \bigwedge_{0 \le k \le n} p_k \approx q_k).$$

P r o o f. Assume that

(1) $$p = \rho(p_0,p_1,\dots,p_n),$$

(2) $$q = \rho(q_0,q_1,\dots,q_n)$$

and additionally that

(1.1) $$p \approx q.$$

Then by Axiom II.10 from formulas (1) and (1.1) we obtain

(1.2) $$q = \rho(r_0, r_1, \ldots, r_n)$$

and

(1.3) $$\bigwedge_{0 \leq k \leq n} r_k \approx p_k.$$

Since by Axiom II.9 the function ρ is a one-to-one function from (2) and (1.2) we obtain

(1.4) $$\bigwedge_{0 \leq k \leq n} q_k = r_k,$$

which by (1.3) yields

(1.5) $$\bigwedge_{0 \leq k \leq n} q_k \approx p_k.$$

Assume now that

(2.1) $$\bigwedge_{0 \leq k \leq n} q_k \approx p_k.$$

Then by Axiom II.10 and formulas (1), (2) and (2.1) we have

(2.2) $$q \approx p$$

whence by Axiom I.1b we obtain

(2.3) $$p \approx q.$$

Since the assumption (1.1) yields (1.5) and the assumption (2.1) yields (2.3) the proof of the conclusion is concluded.

THEOREM II.2

$$S = \bigcap \{ X \mid \delta(X) \}.$$

The proof of Theorem II.2 is based on the two inclusions:

THEOREM II.2a

$$S \subseteq \bigcap \{ X \mid \delta(X) \},$$

THEOREM II.2b

$$\bigcap \{ X \mid \delta(X) \} \subseteq S.$$

P r o o f of II.2a. In accordance with Definition II.4c it is to be demonstrated that

$$(\alpha) \qquad r \in \bigcup_{k=0}^{\infty} {}^{k}S \Rightarrow \bigwedge_{X} (\delta(X) \Rightarrow r \in X).$$

It suffices to demonstrate that the following implication holds:

$$(\alpha') \qquad r \in {}^{k}S \Rightarrow \bigwedge_{X} (\delta(X) \Rightarrow r \in X).$$

The proof of (α') is by induction. For $k=0$, $r \in {}^{0}S$ and, by Definition II.4a, $r \in E_S$, whence the consequent of (α') follows self-evidently by Definition II.5. Let, therefore, the implication be true for $k=l$. We shall prove that it is also true for $k=l+1$. Assume accordingly that $r \in {}^{l+1}S$. By Definition II.4b and Convention II.3c we have

$$(1) \qquad r \in {}^{l}S$$

or

$$(2) \qquad n_1 \geq 1 \wedge r_0, r_1, \ldots, r_{n_1} \in {}^{l}S \wedge r = \rho(r_0, r_1, \ldots, r_{n_1}) \wedge \\ \wedge\, C^{n_1+1}(c, c_1, \ldots, c_{n_1}, c_0).$$

In case (1) (α') is true by the inductive assumption. In case (2) we have

(3) $$n_1 \geq 1 \wedge r_0, r_1, \ldots, r_{n_1} \in {}^{l}S$$

and

(4) $$r = \rho(r_0, r_1, \ldots, r_{n_1}) \wedge c^{n_1+1}(c, c_1, \ldots, c_{n_1}, c_0)$$

and, by (3) and the inductive assumption, we also have

(5) $$\bigwedge_X (\delta(X) \Rightarrow r_0, r_1, \ldots, r_{n_1} \in X) \wedge n_1 \geq 1.$$

Thus when X is any set that satisfies the condition $\delta(X)$, we obtain from (5) that $r_0, r_1, \ldots, r_{n_1} \in X$, and in view of (4) we obtain by Definition II.5 that $r \in X$.

In both cases, (1) and (2), it follows from the inductive assumption that (α') is true for $k = l+1$. This statement concludes the proof of (α').

P r o o f of II.2b. This theorem is equivalent to the implication converse to (α), that is to the implication

(β) $$\bigwedge_X (\delta(X) \Rightarrow r \in X) \Rightarrow r \in \bigcup_{k=0}^{\infty} {}^{k}S.$$

Assume that the antecedent of (β) is true so that, in accordance with Definition II.5, r is in any set X that satisfies the following conditions:

(1) $$E_S \subseteq X,$$

(2) $$\bigwedge_\rho \bigwedge_{n \geq 1} \bigwedge_{\rho_0, \rho_1, \ldots, \rho_n \in X} (\rho = \rho(\rho_0, \rho_1, \ldots, \rho_n) \wedge$$
$$\wedge\, c^{n+1}(\alpha, \alpha_1, \ldots, \alpha_n, \alpha_0) \Rightarrow \rho \in X).$$

In order to demonstrate that $r \in \bigcup_{k=0}^{\infty} {}^{k}S$ it suffices to prove that the set $\bigcup_{k=0}^{\infty} {}^{k}S$ satisfies conditions (1) and (2). In fact, it satisfies condition (1), i.e., $E_s \subseteq \bigcup_{k=0}^{\infty} {}^{k}S$, because by Definition II.4a ${}^{0}S=E_s$. To demonstrate that it satisfies condition (2) let us assume that

$$(3) \qquad n \geq 1 \wedge p_0, p_1, \ldots, p_n \in \bigcup_{k=0}^{\infty} {}^{k}S$$

and

$$(4) \qquad p = \rho(p_0, p_1, \ldots, p_n) \wedge c^{n+1}(\alpha, \alpha_1, \ldots, \alpha_n, \alpha_0).$$

By (3) we can state that there is the greatest natural number l out of the numbers which determine the order of the expressions $p_0, p_1, \ldots, p_n$ $(n \geq 1)$. Thus by Lemma II.1 we have

$$(5) \qquad n \geq 1 \wedge p_0, p_1, \ldots, p_n \in {}^{l}S.$$

By Definition II.4b formulas (5) and (4) yield $p \in {}^{l+1}S$, whence $p \in \bigcup_{k=0}^{\infty} {}^{k}S$. The set $\bigcup_{k=0}^{\infty} {}^{k}S$ satisfies the required condition (2). The implication (β) is accordingly true.

LEMMA II.3

$$p \in {}^{k}S \wedge q \approx p \Rightarrow q \in {}^{k}S.$$

P r o o f. The proof of this lemma is by induction. By Definition II.4a and Conclusion II.4a the lemma is true for $k=0$. Let us assume its being true for $k=l$ and demonstrate its also being true for $k=l+1$. Let accordingly

$$(1) \qquad p \in {}^{l+1}S$$

and

$$(2) \qquad q \approx p.$$

By Definition II.4b it follows from (1) that

(3) $$p \in {}^{l}S$$

or

(4) $$\bigvee_{n\geq 1} \bigvee_{p_0, p_1, \ldots, p_n \in {}^{l}S} (p = \rho(p_0, p_1, \ldots, p_n) \wedge$$

$$\wedge\ c^{n+1}(\alpha, \alpha_1, \ldots, \alpha_n, \alpha_0)).$$

Now for (3) the lemma follows immediately from the inductive assumption in view of (2) and Definition II.4b. For (4), in view of Convention II.3c, we have

(5) $$n_1 \geq 1 \wedge r_0, r_1, \ldots, r_{n_1} \in {}^{l}S,$$

(6) $$p = \rho(r_0, r_1, \ldots, r_{n_1}),$$

(7) $$c^{n+1}(\alpha, c_1, \ldots, c_{n_1}, c_0).$$

By Axiom II.10, (6) and (2) yield

(8) $$q = \rho(s_0, s_1, \ldots, s_{n_1})$$

and

(9) $$\bigwedge_{0\leq k\leq n_1} s_k \approx r_k.$$

Further, (5) and (9) yield, by the inductive assumption,

(10) $$n_1 \geq 1 \wedge s_0, s_1, \ldots, s_{n_1} \in {}^{l}S.$$

By Convention II.3c, it follows from (10), (5) and (9), Definition II.4c and Conclusions II.2b and II.6a that

(11) $$\bigwedge_{0\leq k\leq n_1} d_k \approx c_k.$$

Since at the same time

(12) $$b \approx a$$

by (1), Lemma II.2, (2), Axiom II.8 and Convention II.3c, it follows from (7), (11), (12), Axiom I.1b and Conclusion I.11 that

(13) $$C^{n+1}(b,d_1,\ldots,d_n,d_0).$$

In view of (10), (8) and (13) we find, by Definition II.4b, that $q\in{}^{l+1}S$. Hence the lemma under consideration is, under the inductive assumption, true for $k=l+1$ also in case (4). This concludes the inductive proof of the lemma.

LEMMA III.1

$$^{n}S \subseteq E \subseteq D(\iota).$$

P r o o f. In view of Conclusion II.5b we have

(1) $$E \subseteq D(\iota),$$

so that it suffices to demonstrate that

(2) $$^{n}S \subseteq E$$

holds.

The proof of (2) is by induction. For $n=0$ (2) follows immediately from Definitions III.2a and II.3c. Let us assume inductively that it is true for $n=k$. We shall demonstrate that it is true for $n=k+1$. Let us assume for that purpose that $p\in{}^{k+1}S$. Then, by Definition III.2b, we have one of the three cases (1°,2°,3°), listed in the verbal formulation of that definition. In case 1°, when $p\in{}^{k}S$, formula (2) holds in accordance with the inductive assumption. In case 2°, p is a compound expression consisting of elements that belong to ${}^{k}S$; in accordance with the inductive assumption they are in E, and by (1) they are in $D(\iota)$. Hence, by Axiom II.9, $p \in D^{-1}(\rho)$; thus p is in E_{c} (Definition II.3b) and in E (Definition II.3c). Finally, in case 3° p is an operator expression, and hence, by Definition III.1*, Conclusions III.1a,b and formula (1) it is a compound expression consisting of an operator, its indexical variable and its scope, all in $D(\iota)$. Thus in case 3°, by Axiom II.9, $p \in D^{-1}(\rho)$ and hence, by Definitions II.3b,c, $p \in E$. Thus it follows from the inductive assumption that (2) is true for $n=k+1$, which was to be demonstrated.

LEMMA III.2

$$p \in {}^{k}S \wedge q \approx p \Rightarrow q \in {}^{k}S.$$

P r o o f. The proof of this lemma is by induction. For $k=0$ the lemma is true by Definition III.2a and Conclusion II.4a. We assume that the lemma is true for $k=l$ and shall demonstrate its being true for $k=l+1$. For that purpose we assume additionally that

(1) $$p \in {}^{l+1}S$$

and

(2) $$q \approx p.$$

By Definition III.2b it follows from (1) that one of the three cases holds:

1° $$p \in {}^{l}S,$$

2° $\bigvee_{n\geq 1} \bigvee_{p_0,p_1,\ldots,p_n \in {}^{l}S} (p = \rho(p_0,p_1,\ldots,p_n) \wedge p_0 \notin O \wedge$

$\wedge\ c^{n+1}(\alpha,\alpha_1,\ldots,\alpha_n,\alpha_0)),$

3° $\bigvee_{p_0,p_1,p_2,p_3} (p = \rho^{\omega}(p_0,p_1,p_2) \wedge p_2 \in {}^{l}S \wedge p_3(fv)p_2 \wedge$

$\wedge\ p_3 \approx p_1 \wedge c^{3}(\alpha,\alpha_1,\alpha_2,\alpha_0)).$

In case 1°, $q \in {}^{l+1}S$ by the inductive assumption, formula (2), and Definition III.2b. In case 2°, the proof of the fact that $q \in {}^{l+1}S$ is quite analogical to the proof of the fact that $q \in {}^{l+1}S$ in the inductive proof of Lemma II.3 in case (4). It suffices merely, instead of Definition II.4c, Conclusion II.6a, Lemma II.2 and Definition II.4b, to refer, respectively, to Definition III.2c, Conclusion III.6, Lemma III.1 and Definition III.2b and to make use of Axiom III.2.

Note that on the strength of the inductive assumption the expression q, equiform with the expression p that satisfies condition 2°, i.e., an expression which is in the set $S^{n\omega}$ (see Definition III.3), is also in $S^{n\omega}$.

Finally we have to consider case 3°. In that case there is a sequence of expressions p_0,p_1,p_2,p_3 such that

(3) $p = \rho^{\omega}(p_0,p_1,p_2),$

(4) $p_2 \in {}^{l}S,$

(5) $p_3(fv)p_2,$

(6) $p_3 \approx p_1,$

and

(7) $c^{3}(\alpha,\alpha_1,\alpha_2,\alpha_0).$

It follows from Theorem III.1 and formulas (3) and (2) that there exist expressions q_0, q_1, q_2 such that

(8) $$q = \rho^{\omega}(q_0, q_1, q_2)$$

and

(9) $$q_0 \approx p_0 \wedge q_1 \approx p_1 \wedge q_2 \approx p_2.$$

It follows from the inductive assumption and formulas (4) and (9) that

(10) $$q_2 \in {}^{l}S.$$

It follows in turn from Axiom I.1a, Conclusion III.4, and formulas (5) and (9) that there is a q_3 such that

(11) $$q_3(fv)q_2$$

and

(12) $$q_3 \approx p_3.$$

By Axioms I.1c,b we obtain from (12), (6) and (9) the formula

(13) $$q_3 \approx q_1.$$

It follows from (1) and Lemma III.1 that $p \in D(\iota)$. This and formula (2) yield, by Axiom II.8 and Convention II.3c,

(14) $$b \approx a.$$

Since, by Convention II.3a, p_0, p_1, p_2 are expressions (which is also guaranteed by Definitions III.1* and II.3c and Axiom III.1) they belong to the set $D(\iota)$ (Conclusion II.5b). By making use again of Axiom II.8 and Convention II.3c we arrive, on the strength of (9), at the formula of the form

(15) $$\bigwedge_{0\leq k\leq n} b_k \approx a_k.$$

From formulas (7), (14) and (15) we obtain, by Axiom I.1b and Conclusion I.11,

(16) $$c^3(b,b_1,b_2,b_0).$$

Thus in view of (8), (10), (11), (13) and (16) we have

(17) $$\bigvee_{q_0,q_1,q_2,q_3} (q = \rho^{\omega}(q_0,q_1,q_2) \wedge q_2 \in {}^{l}S \wedge$$
$$\wedge\ q_3(fv)q_2 \wedge q_3 \approx q_1 \wedge c^3(b,b_1,b_2,b_0))$$

and in accordance with Definition III.2b, in view of (17), we obtain $q\in{}^{l+1}S$. Thus also in case 3° the truth of the lemma for $k=l+1$ follows from the assumption of its being true for $k=l$. The lemma is accordingly proved.

Note also that it follows from the inductive assumption that in case 3°, when $p\in S^{\omega}$ by Definition III.4, an expression q which is equiform with p is also in S^{ω} because formula (17) holds.

THEOREM III.6a (I fttsc)

$$r(p/q)s \wedge r \underset{c}{\equiv} s \Rightarrow p \underset{c}{\equiv} q.$$

P r o o f. The theorem follows immediately from Definition III.8d and

LEMMA III.3

$$r(p/q)^n s \wedge r \underset{c}{\equiv} s \Rightarrow p \underset{c}{\equiv} q,$$

which will be proved by induction.

For $n=0$ Lemma III.3 is true on the strength of Definition III.8a, Theorem III.3, Convention II.3c and Conclusions III.8,

II.12, and II.13b,c.

Assume that for n=1

(1a) $$r(p/q)^1 s$$

and

(1b) $$r \underset{c}{\equiv} s.$$

We have to demonstrate that $p \underset{c}{\equiv} q$. In accordance with Definition III.8b and assumption (1a) two cases are possible; they are the cases 1° and 2°, described verbally after Definition III.8. Note that in case 1°, when r and s are expressions from the set $S^{n\omega}$, the proof of the lemma now under consideration is quite analogical to the proof of Lemma II.4 which substantiates I **fttsc** in **TSCL**. It suffices only, in the successive steps of the proof, instead of Definitions II.14b and II.4a,b,c and Conclusions II.6a and II.7, which are valid in **TSCL**, to refer to their analoga in **TSCω-L**, that is, respectively, to Definitions III.8b(1°), III.2a, III.3, III.2c, and Conclusions III.6 and III.8. Let us, therefore, consider case 2°. We have then

(2) $$r,s \in S^{\omega}$$

and there are s_0,s_1,s_2 and r_0,r_1,r_2 such that

(3) $$s \equiv \rho^{\omega}(s_0,s_1,s_2),$$

(4) $$r \equiv \rho^{\omega}(r_0,r_1,r_0)$$

and one of the conditions (a), (b), (c), formulated in the verbal comment to Definition III.8b holds. Before we proceed to analyse each of those conditions we have to note that it follows from (3), (4), Definition III.1* and Conclusions III.7a,b that s_0, r_0 and s_1, r_1 are well-formed expressions. Since s and r, given by (3) and (4), have their expansions determined unambiguously - which follows from Defintion III.1*, formula (2), Conclusion III.5b, Definition II.3b and Axiom II.9 - the expressions s_2 and r_2 are well-formed, too, in view of (2) and Definitions III.4 and III.2c. We accordingly have

(5) $$s_0, s_1, s_2, r_0, r_1, r_2 \in S.$$

Note also that since the expansions of s and r are determined unambiguously, it follows from (2)-(4) and Definition III.4 that the conditions $\mathfrak{m}_s$ and $\mathfrak{m}_r$ are satisfied. In accordance with Convention II.3c these conditions take on the form, respectively,

(6) $$c^3(d, d_1, d_2, d_0)$$

and

(7) $$c^3(c, c_1, c_2, c_0).$$

Note also that

(8) $$d \approx c$$

follows from the assumption (1b) and Conclusions II.12 and II.13b.

We now proceed to analyse conditions (a), (b) and (c). We assume first that the following condition is satisfied:

(a) $$q_0 \approx s_0 \wedge p \approx r_0 \wedge s_1 \approx r_1 \wedge s_2 \approx r_2.$$

Then it follows from (5), Conclusions III.8 and Convention II.3c that $d_1 \approx c_1$ and $d_2 \approx c_2$. This yields, by (8), (6) and (7) and Theorem I.17, $d_0 \approx c_0$, and since $q \approx s_0$ and $p \approx r_0$, we have $p \underset{c}{=} q$ in view of (5), Theorem III.3, Convention II.3c and Conclusions III.8, II.12 and II.13b,c. Thus the lemma is true for $n=1$ if condition (a) is satisfied.

Let us consider the next condition

(b) $$q \approx s_1 \wedge p \approx r_1 \wedge \bigwedge_{0 \leq i \leq 2} s_i \approx r_i .$$

It can easily be seen that (b) implies $s_1 \approx r_1$. Thus, if (b) holds, by (5), Theorem III.3, Convention II.3c and Conclusion III.8 we have $b \approx d_1$, $a \approx c_1$ and $d \approx c_1$. This yields immediately $p \underset{c}{=} q$ in view of Conclusions II.12 and II.13b,c. Thus the lemma is true for $n=1$ if condition (b) is satisfied.

We have still to demonstrate that $p \underset{c}{=} q$ if the following condition holds:

(c) $$q \approx s_2 \wedge p \approx r_2 \wedge s_0 \approx r_0 \wedge s_1 \approx r_1 .$$

Now let (c) hold. Then in view of (5), Conclusion III.8 and Convention II.3c we have $d_1 \approx c_1$ and $d_0 \approx c_0$, so that we obtain $d_2 \approx c_2$ in view of (8), (6), (7) and Theorem I.18. Since $q \approx s_2$ and $p \approx r_2$, we have $p \underset{c}{=} q$ in view of (5), Theorem III.3, Convention II.3c, and Conclusions III.8, II.12 and II.13b.

Thus Lemma III.3 is true for $n=1$.

The proof of the fact that the assumption of that lemma being true for $n=k$ ($k>0$) results in its being true for $n=k+1$ is quite similar to the proof of Lemma II.4 for $n=k+1$, on the assumption of that lemma being true for $n=k$. Instead of Definition II.14c it suffices to refer to Definition III.8c.

This concludes the inductive proof of Lemma III.3.

THEOREM IV.4

$$\overline{W} = \bigcap \{ \overline{X} \mid \overline{\lambda}(\overline{X}) \}.$$

To prove this theorem it suffices to prove two lemmata:

LEMMA IV.1

$$\overline{s} \in \overline{W} \Rightarrow \bigwedge_{\overline{X}} (\overline{V} \subseteq \overline{X} \wedge \bigwedge_{\overline{p},\overline{q},\overline{r}} (\overline{p},\overline{q} \in \overline{X} \wedge \overline{c}(\overline{p},\overline{q},\overline{r}) \Rightarrow \overline{r} \in \overline{X}) \Rightarrow \overline{s} \in \overline{X})$$

and its converse

LEMMA IV.2

$$\bigwedge_{\overline{X}} (\overline{V} \subseteq \overline{X} \wedge \bigwedge_{\overline{p},\overline{q},\overline{r}} (\overline{p},\overline{q} \in \overline{X} \wedge \overline{c}(\overline{p},\overline{q},\overline{r}) \Rightarrow \overline{r} \in \overline{X}) \Rightarrow \overline{s} \in \overline{X}) \Rightarrow \overline{s} \in \overline{W}.$$

P r o o f of Lemma IV.1. Assume that

(1) $$\overline{s} \in \overline{W}$$

and additionally that

(2) $$\overline{V} \subseteq \overline{X}$$

and

(3) $$\bigwedge_{\overline{p},\overline{q},\overline{r}} (\overline{p},\overline{q} \in \overline{X} \wedge \overline{c}(\overline{p},\overline{q},\overline{r}) \Rightarrow \overline{r} \in \overline{X}).$$

We have to demonstrate that $\overline{s} \in \overline{X}$.

It follows from (1) and Conclusion IV.4 that there is a label token s_1 such that

(4) $$s_1 \in W$$

and

(5) $$\overline{s} = [s_1].$$

From formula (4) we obtain, by Definitions I.1 and I.1a,

(6) $$\bigwedge_{X} (V \subseteq X \wedge \bigwedge_{p,q,r} (p,q \in X \wedge c(p,q,r) \Rightarrow r \in X) \Rightarrow s_1 \in X).$$

Assume that

(7) $$X_0 = \{ p \mid [p] \in \overline{X} \}.$$

It can easily be seen that X_0 satisfies the condition $\lambda(X_0)$ (see Definition I.1a).

In fact, if $p \in V$, then $[p] \in \overline{V}$ in view of Conclusion IV.3 and then it follows from (2) and (7) that $p \in X_0$.

We thus have

(8) $$V \subseteq X_0.$$

If, on the other hand, we assume that $p,q \in X_0$ and $c(p,q,r)$, then by (7) $[p]$ and $[q]$ belong to $\overline{X}$ and by Definition IV.2 we have $\overline{c}([p],[q],[r])$. By applying formula (3) (see Convention IV.2a and Conclusion IV.1) we obtain $[r] \in \overline{X}$ so that, in view of (7), $r \in X_0$. Hence

(9) $$\bigwedge_{p,q,r} (p,q \in X_0 \wedge c(p,q,r) \Rightarrow r \in X_0).$$

From (6), (8) and (9) we obtain $s_1 \in X_0$, whence in view of (7) and (5) we have $\overline{s} \in \overline{X}$. Lemma IV.1 has thus been proved.

P r o o f of Lemma IV.2. Assume that

(1) $$\bigwedge_{\overline{X}} (\overline{V} \subseteq \overline{X} \wedge \bigwedge_{\overline{p},\overline{q},\overline{r}} (\overline{p},\overline{q} \in \overline{X} \wedge \overline{c}(\overline{p},\overline{q},\overline{r}) \Rightarrow \overline{r} \in \overline{X}) \Rightarrow \overline{s} \in \overline{X}).$$

In order to demonstrate that $\overline{s} \in \overline{W}$ it suffices to show that $\overline{W}$ satisfies the condition $\overline{\lambda}(\overline{X})$ (see Definition IV.5).

In fact it follows from Conclusions IV.3, IV.4 and I.3a that

(2) $$\overline{V} \subseteq \overline{W}.$$

It can also be proved that

(3) $$\bigwedge_{\overline{p},\overline{q},\overline{r}} (\overline{p},\overline{q} \in \overline{W} \wedge \overline{c}(\overline{p},\overline{q},\overline{r}) \Rightarrow \overline{r} \in \overline{W}).$$

To do so let us assume that $\overline{p},\overline{q} \in \overline{W}$ and $\overline{c}(\overline{p},\overline{q},\overline{r})$. Then it follows from Conclusion IV.4 that there are word tokens p_1 and q_1 such that $\bar{p}=[p_1]$ and $\bar{q}=[q_1]$. Further it follows from Convention IV.2a and Conclusions IV.1 and IV.2 that there is a label token r_1 such that $\bar{r}=[r_1]$ and $c(p_1,q_1,r_1)$. Since p_1,q_1 are word tokens, r_1 is also a word token in view of Conclusion I.3b, and hence $\bar{r}\in\bar{W}$ by Conclusion IV.4. Formula (3) has thus been proved.

Now formulas (1), (2) and (3) imply that $\bar{s}\in\bar{W}$, what was to be demonstrated.

THEOREM IV.22b

$$\bar{p} \in {}^{k+1}\bar{S} \Leftrightarrow \bar{p} \in {}^{k}\bar{S} \vee \bigvee_{n\geq 1} \bigvee_{\bar{p}_0,\bar{p}_1,\ldots,\bar{p}_n\in {}^{k}\bar{S}} (\bar{p} = \bar{p}(\bar{p}_0,\bar{p}_1,\ldots,\bar{p}_n) \wedge \bar{c}^{n+1}(\bar{a},\bar{a}_1,\ldots,\bar{a}_n,\bar{a}_0)).$$

P r o o f. To prove the direct implication we assume that

(1) $$\bar{p} \in {}^{k+1}\bar{S}.$$

We shall demonstrate that

(v) $$\bar{p} \in {}^{k}\bar{S} \vee \bigvee_{n\geq 1} \bigvee_{\bar{p}_0,\bar{p}_1,\ldots,\bar{p}_n\in {}^{k}\bar{S}} (\bar{p} = \bar{p}(\bar{p}_0,\bar{p}_1,\ldots,\bar{p}_n) \wedge \bar{c}^{n+1}(\bar{a},\bar{a}_1,\ldots,\bar{a}_n,\bar{a}_0)).$$

By Conclusion IV.20 it follows from (1) that there is a q_1 such that

(2) $$\bar{p} = [q_1]$$

and

(3) $$q_1 \in {}^{k+1}S.$$

It follows from (3) and Definition II.4b that there may be two cases:

(i) $$q_1 \in {}^k S$$

or

(ii) $$\bigvee_{n\geq 1} \bigvee_{p_0, p_1, \ldots, p_n \in {}^k S} (q_1 = \rho(p_0, p_1, \ldots, p_n) \wedge$$

$$\wedge\ c^{n+1}(b_1, a_1, \ldots, a_n, a_0)).$$

In case (i) $\bar{p} = [q_1]$ in accordance with (2) and hence $\bar{p} \in {}^k\bar{S}$ by Conclusion IV.20 and (v) is true. Consider then case (ii). Now (2), (ii) and Convention II.3c yield

(4) $$r_0, r_1, \ldots, r_{n_1} \in {}^k S \wedge q_1 = \rho(r_0, r_1, \ldots, r_{n_1}) \wedge$$

$$\wedge\ c^{n+1}(b_1, c_1, \ldots, c_{n_1}, c_0) \wedge n_1 \geq 1,$$

(5) $$\bar{p} = [\rho(r_0, r_1, \ldots, r_{n_1})].$$

Assume that

(6) $$\bar{r}_0 = [r_0] \wedge \bar{r}_1 = [r_1] \wedge \ldots \wedge \bar{r}_{n_1} = [r_{n_1}].$$

It follows from formulas (6), (2), (4), Lemma II.2, Theorem IV.16a, and Conventions II.3c and IV.3c that

(7) $$\bar{c}_0 = [c_0] \wedge \bar{c}_1 = [c_1] \wedge \ldots \wedge \bar{c}_{n_1} = [c_{n_1}] \wedge \bar{a} = [b_1].$$

In view of Conclusion IV.20 and formulas (4) and (6) we see that

(8) $$\bar{r}_0, \bar{r}_1, \ldots, \bar{r}_{n_1} \in {}^k\bar{S}.$$

Since in accordance with Lemma II.2 and formula (4) $n_1 \geq 1$ and $r_1, r_2, \ldots, r_{n_1} \in D(\iota)$, it follows from (6) and (5) and Theorem IV.17a that

(9) $$\bar{\rho} = \bar{\rho}(\bar{r}_0,\bar{r}_1,\ldots,\bar{r}_{n_1}).$$

Further, in view of (7) and (4) and Conclusion IV.6 we obtain

(10) $$\bar{c}^{n+1}(\bar{\alpha},\bar{c}_1,\ldots,\bar{c}_{n_1},\bar{c}_0) \wedge n_1 \geq 1.$$

The truth of (v) in the second of the two possible cases follows from formulas (8)-(10).

Now let (v) hold. It is to be demonstrated that $\bar{\rho}\in{}^{k+1}\bar{S}$. If the first constituent of the disjunction (v) holds, then $\rho\in{}^{k+1}\bar{S}$ follows immediately from Conclusion IV.20 and Lemma II.1. Let therefore the second constituent of the disjunction (v) hold. Then there is an $n_1 \geq 1$ and the following formulas are valid:

(1) $$\bar{s}_0,\bar{s}_1,\ldots,\bar{s}_{n_1} \in {}^{k}\bar{S},$$

(2) $$\bar{\rho} = \bar{\rho}(\bar{s}_0,\bar{s}_1,\ldots,\bar{s}_{n_1})$$

and

(3) $$\bar{c}^{n+1}(\bar{\alpha},\bar{d}_1,\ldots,\bar{d}_{n_1},\bar{d}_0).$$

In view of Conclusion IV.20 and formula (1) we can state that there are $s_0,s_1,\ldots,s_{n_1}$ such that

(4) $$s_0,s_1,\ldots,s_{n_1} \in {}^{k}S$$

and

(5) $$\bar{s}_0 = [s_0] \wedge \bar{s}_1 = [s_1] \wedge \ldots \wedge \bar{s}_{n_1} = [s_{n_1}].$$

Since $n_1 \geq 1$ and, by Lemma II.2 and formula (4), $s_0,s_1,\ldots,s_{n_1} \in D(\iota)$, we find, by referring to (2) and (5) and Theorem IV.17a, that

(6) $$\bar{p} = [\rho(s_0,s_1,\dots,s_{n_1})].$$

The expression $\rho(s_0,s_1,\dots,s_{n_1})$ is in $D(\iota)$, which follows from Axiom II.9. By applying Convention IV.3c and referring to Theorem IV.16a we obtain from (6)

(7) $$\bar{\alpha} = [\iota(\rho(s_0,s_1,\dots,s_{n_1}))].$$

From (5) we obtain immediately, by Theorem IV.16a, Conventions IV.3c and II.3c,

(8) $$\bar{d}_0 = [d_0] \wedge \bar{d}_1 = [d_1] \wedge \dots \wedge \bar{d}_{n_1} = [d_{n_1}].$$

By referring to Conclusion IV.6, formulas (7), (8) and (3) and the fact that $n_1+1>1$ we obtain

(9) $$c^{n_1+1}(\iota(\rho(s_0,s_1,\dots,s_{n_1})),d_1,\dots,d_{n_1},d_0).$$

In view of Definition II.4b and formulas (4) and (9) we find that $\rho(s_0,s_1,\dots,s_{n_1})\in{}^{k+1}S$, whence, in view of formula (6) and Conclusion IV.20, we obtain $\bar{p}\in{}^{k+1}\bar{S}$, which was to be demonstrated.

THEOREM IV.26b

$$\bar{r}(\overline{p/q})^1\bar{s} \Leftrightarrow \bar{s},\bar{r} \in \bar{S} \wedge \bigvee_{n\geq 1} \bigvee_{\substack{\bar{s}_0,\bar{s}_1,\dots,\bar{s}_n\in\bar{S} \\ \bar{r}_0,\bar{r}_1,\dots,\bar{r}_n\in\bar{S}}} (\bar{s} = \bar{\rho}(\bar{s}_0,\bar{s}_1,\dots,\bar{s}_n) \wedge \bar{r} = \bar{\rho}(\bar{r}_0,\bar{r}_1,\dots,\bar{r}_n) \wedge$$

$$\wedge \bigvee_{0\leq j\leq n} (\bar{q} = \bar{s}_j \wedge \bar{p} = \bar{r}_j \wedge \bigwedge_{\substack{k\neq j \\ 0\leq k\leq n}} \bar{s}_k = \bar{r}_k)),$$

P r o o f. The following list of equivalences results from

Definitions IV.26a and II.14b, Conclusion II.6a, Theorem IV.17a and Conclusions IV.19, IV.18 and II.7:

$$\overline{r}(\overline{\rho}/\overline{q})^{1}\overline{s} \Leftrightarrow \bigvee_{\rho,q,r,s}(\overline{\rho}=[\rho]\wedge\overline{q}=[q]\wedge\overline{r}=[r]\wedge\overline{s}=[s]\wedge$$

$$\wedge\, r(\rho/q)^{1}s) \Leftrightarrow \bigvee_{\rho,q,r,s}(\overline{\rho}=[\rho]\wedge\overline{q}=[q]\wedge\overline{r}=[r]\wedge\overline{s}=[s]\wedge$$

$$\wedge\, s,r\in S\wedge\bigvee_{n\geq 1}\ \bigvee_{\substack{s_0,s_1,\ldots,s_n\in S\\ r_0,r_1,\ldots,r_n\in S}}(s=\rho(s_0,s_1,\ldots,s_n)\wedge$$

$$\wedge\, r=\rho(r_0,r_1,\ldots,r_n)\wedge\bigvee_{0\leq j\leq n}(q\approx s_j\wedge\rho\approx r_j\wedge$$

$$\wedge\bigwedge_{\substack{k\neq j\\ 0\leq k\leq n}} s_k\approx r_k))) \Leftrightarrow \bigvee_{\rho,q,r,s}\ \bigvee_{n\geq 1}\ \bigvee_{\substack{s_0,s_1,\ldots,s_n\in S\\ r_0,r_1,\ldots,r_n\in S}}\ \bigvee_{0\leq j\leq n}(\overline{\rho}=[\rho]\wedge$$

$$\wedge\,\overline{q}=[q]\wedge\overline{r}=[r]\wedge\overline{s}=[s]\wedge s,r\in S\wedge s=\rho(s_0,s_1,\ldots,s_n)\wedge$$

$$\wedge\, r=\rho(r_0,r_1,\ldots,r_n)\wedge q\approx s_j\wedge\rho\approx r_j\wedge\bigwedge_{\substack{k\neq j\\ 0\leq k\leq n}} s_k\approx r_k) \Leftrightarrow$$

$$\Leftrightarrow \bigvee_{n>1}\ \bigvee_{\substack{s_0,s_1,\ldots,s_n\in S\\ r_0,r_1,\ldots,r_n\in S}}\ \bigvee_{0\leq j\leq n}(\overline{s}\in\overline{S}\wedge\overline{r}\in\overline{S}\wedge$$

$$\wedge\,\overline{s}=\overline{\rho}([s_0],[s_1],\ldots,[s_n])\wedge\overline{r}=\overline{\rho}([r_0],[r_1],\ldots,[r_n])\wedge$$

$$\wedge\,\overline{q}=[s_j]\wedge\overline{\rho}=[r_j]\wedge\bigwedge_{\substack{k\neq j\\ 0\leq k\leq n}}[s_k]=[r_k] \Leftrightarrow$$

$$\Leftrightarrow \bigvee_{n\geq 1}\ \bigvee_{\substack{\overline{s}_0,\overline{s}_1,\ldots,\overline{s}_n\in\overline{S}\\ \overline{r}_0,\overline{r}_1,\ldots,\overline{r}_n\in\overline{S}}}(\overline{s},\overline{r}\in\overline{S}\wedge\overline{s}=\overline{\rho}(\overline{s}_0,\overline{s}_1,\ldots,\overline{s}_n)\wedge$$

$$\wedge\,\overline{r}=\overline{\rho}(\overline{r}_0,\overline{r}_1,\ldots,\overline{r}_n)\wedge\bigvee_{0\leq j\leq n}(\overline{q}=\overline{s}_j\wedge\overline{\rho}=\overline{r}_j\wedge$$

$$\wedge\bigwedge_{\substack{k\neq j\\ 0\leq k\leq n}}\overline{s}_k=\overline{r}_k)).$$

According to the above list of equivalence both sides of the equivalence under consideration are equivalent.

THEOREM IV.26c

$$k > 0 \Rightarrow (\bar{r}(\bar{p}/\bar{q})^{k+1}\bar{s} \Leftrightarrow \bigvee_{\bar{t},\bar{v}} (\bar{r}(\bar{v}/\bar{t})^{k}\bar{s} \wedge \bar{v}(\bar{p}/\bar{q})^{1}\bar{t})).$$

P r o o f. Let $k>0$. We refer to Definitions IV.26a and II.14c, Conclusion IV.1 and the relationship $r_1 \approx r_2 \wedge s_1 \approx s_2 \wedge \wedge r_1(p/q)^1 s_1 \Rightarrow r_2(p/q)^1 s_2$ (which follows from Definition II.14b and Axioms II.10 and I.1c) to prove the following list of equivalences and implications:

$$\bar{r}(\bar{p}/\bar{q})^{k+1}\bar{s} \Leftrightarrow \bigvee_{p,q,r,s} (\bar{p} = [p] \wedge \bar{q} = [q] \wedge \bar{r} = [r] \wedge \bar{s} = [s] \wedge$$

$$\wedge\, r(p/q)^{k+1}s \Leftrightarrow \bigvee_{p,q,r,s} (\bar{p} = [p] \wedge \bar{q} = [q] \wedge \bar{r} = [r] \wedge \bar{s} = [s] \wedge$$

$$\wedge \bigvee_{t,v} (r(v/t)^{k}s \wedge v(p/q)^{1}t)) \Leftrightarrow \bigvee_{p,q,r,s,t,v} (\bar{p} = [p] \wedge$$

$$\bar{q} = [q] \wedge \bar{r} = [r] \wedge \bar{s} = [s] \wedge r(v/t)^{k}s \wedge v(p/q)^{1}t) \Rightarrow$$

$$\Rightarrow \bar{r}([v]/[t])^{k}\bar{s} \wedge [v](\bar{p}/\bar{q})^{1}[t] \Rightarrow \bigvee_{\bar{t},\bar{v}} (\bar{r}(\bar{v}/\bar{t})^{k}\bar{s} \wedge \bar{v}(\bar{p}/\bar{q})^{1}\bar{t}) \Leftrightarrow$$

$$\Leftrightarrow \bigvee_{\bar{t},\bar{v}} (\bigvee_{v,t,r,s} (\bar{v} = [v] \wedge \bar{t} = [t] \wedge \bar{r} = [r] \wedge \bar{s} = [s] \wedge$$

$$\wedge\, r(v/t)^{k}s) \wedge \bigvee_{p,q,v_1,t_1} (\bar{p} = [p] \wedge \bar{q} = [q] \wedge \bar{v} = [v_1] \wedge$$

$$\wedge\, \bar{t} = [t_1] \wedge v_1(p/q)^{1}t_1)) \Rightarrow \bigvee_{p,q,v,v_1,t,t_1,r,s} (\bar{p} = [p] \wedge$$

$$\wedge\ \bar{q} = [q] \wedge \bar{r} = [r] \wedge \bar{s} = [s] \wedge r(v/t)^k s \wedge v_1 \approx v \wedge t_1 \approx t \wedge$$

$$\wedge\ v_1(p/q)^1 t_1 \,) \Rightarrow \bigvee_{p,q,r,s,t,v} (\bar{p} = [p] \wedge \bar{q} = [q] \wedge \bar{r} = [r] \wedge$$

$$\wedge\ \bar{s} = [s] \wedge r(v/t)^k s \wedge v(p/q)^1 t \,).$$

It follows from the above list of equivalences and implications that for the assumption $k>0$ the expression $(\bar{r}(\overline{p/q}))^{k+1}\bar{s}$ implies the expression which is the right side of the equivalence in the theorem under consideration, and that right side in turn implies an expression which is equivalent to the expression $(\bar{r}(\overline{p/q}))^{k+1}\bar{s}$. This concludes the proof of Theorem IV.26c.

THEOREM IV.29

$$\bar{p}(\overline{fv})\bar{q} \Leftrightarrow \bar{p} \in \overline{Vr} \wedge \bar{p} \in \bar{C}_{\bar{q}} \wedge \neg \bigvee_{\bar{r},\bar{r}_0,\bar{r}_1,\bar{r}_2} (\bar{r} = \bar{p}^{\omega}(\bar{r}_0,\bar{r}_1,\bar{r}_2) \wedge$$

$$\wedge\ \bar{r} \in \bar{C}_{\bar{q}} \wedge \bar{p} \in \bar{C}_{\bar{r}_2} \wedge \bar{p} = \bar{r}_1).$$

In the proof of this theorem we avail ourselves of the following lemmata:

LEMMA IV.3

$$\bar{t} \in \bar{C}_{\bar{p}} \wedge \bar{t} = [t] \wedge \bar{p} = [p] \Rightarrow t \in C_p,$$

which follows from Theorem IV.20d, Definition II.12d and

LEMMA IV.3'

$$\bar{t} \in \bar{C}^n_{\bar{p}} \wedge \bar{t} = [t] \wedge \bar{p} = [p] \Rightarrow t \in C^n_p,$$

LEMMA IV.4

$$t \in C_{p} \Rightarrow [t] \in \overline{C}_{[p]},$$

which follows from Theorem IV.20d, Definition II.12d and

LEMMA IV.4'

$$t \in C^{n}_{p} \Rightarrow [t] \in \overline{C}^{n}_{[p]}.$$

Lemma IV.3' follows from Definition IV.21a and Conclusions II.16a,b. Lemma IV.4' follows directly from Definition IV.21a.

P r o o f of the direct implication. Let the antecedent of that implication

(1) $$\overline{p}(\overline{fv})\overline{q}$$

be true. In view of this assumption it follows from Definition IV.29 that there are expressions p_2, p_3 such that

(2) $$\overline{p} = [p_3] \text{ and } \overline{q} = [p_2]$$

and

(3) $$p_3(fv)p_2.$$

In view of (3) we obtain from Definition III.1

(4) $$p_3 \in Vr \wedge p_3 \in C_{p_2}$$

and

(5) $$\neg \bigvee_{r, r_0, r_1, r_2} (r = \rho^{\omega}(r_0, r_1, r_2) \wedge r \in C_{p_2} \wedge \wedge p_3 \in C_{r_2} \wedge p_3 \approx r_1).$$

From (2)-(4) we obtain, by Conclusion IV.24 and Lemma IV.4,

(6) $$\overline{p} \in \overline{Vr} \wedge \overline{p} \in \overline{C}_{\overline{q}}.$$

Assume that

(1.1) $$\bigvee_{r,r_0,r_1,r_2} (\bar{r} = \bar{\rho}^{\omega}(\bar{r}_0,\bar{r}_1,\bar{r}_2) \wedge \bar{r} \in \bar{C}_{\bar{q}} \wedge$$

$$\wedge\ \bar{p} \in \bar{C}_{\bar{r}_2} \wedge \bar{p} = \bar{r}_1).$$

It follows from (1.1) that

(1.2) $$\bar{s} = \bar{\rho}^{\omega}(\bar{s}_0,\bar{s}_1,\bar{s}_2),$$

(1.3) $$\bar{s} \in \bar{C}_{\bar{q}},$$

(1.4) $$\bar{p} \in \bar{C}_{\bar{s}_2},$$

(1.5) $$\bar{p} = \bar{s}_1.$$

In view of (1.2) it follows from Theorem IV.28 that there are s_0,s_1,s_2,s_3 such that

(1.6) $$\bar{s}_0 = [s_0] \wedge \bar{s}_1 = [s_1] \wedge \bar{s}_2 = [s_2] \wedge \bar{s}_3 = [s_3]$$

and

(1.7) $$s_3 = \rho^{\omega}(s_0,s_1,s_2).$$

By Lemma IV.3 it follows from (1.4), (2) and (1.6) that

(1.8) $$s_3 \in C_{p_2},$$

and by Lemma IV.3 it follows from (1.4), (2) and (1.6) that

(1.9) $$p_3 \in C_{s_2}.$$

Further, in view of (1.5), (2) and (1.6) we obtain

(1.10) $$p_3 \approx s_1.$$

But formulas (1.7)-(1.10), taken jointly, contradict formula (5). Hence assumption (1.1) results in a contradiction, and thus the formula

$$\neg \bigvee_{\bar{r},\bar{r}_0,\bar{r}_1,\bar{r}_2} (\bar{r} = \overline{\rho}^{\omega}(\bar{r}_0,\bar{r}_1,\bar{r}_2) \wedge \bar{r} \in \bar{C}_{\bar{q}} \wedge \bar{p} \in \bar{C}_{\bar{r}_2} \wedge \bar{p} = \bar{r}_1)$$

is true. That formula, in view of (6), yields the consequent of the implication under consideration.

P r o o f of the converse implication. Let

(1) $$\bar{p} \in \overline{Vr} \wedge \bar{p} \in \bar{C}_{\bar{q}}$$

and

(2) $$\neg \bigvee_{\bar{r},\bar{r}_0,\bar{r}_1,\bar{r}_2} (\bar{r} = \overline{\rho}^{\omega}(\bar{r}_0,\bar{r}_1,\bar{r}_2) \wedge \bar{r} \in \bar{C}_{\bar{q}} \wedge \bar{p} \in \bar{C}_{\bar{r}_2} \wedge \bar{p} = \bar{r}_1)$$

Assume that

(3) $$\neg (\bar{p}(\overline{fv})\bar{q}).$$

By Definition IV.29, assumption (3) results in

(4) $$\bigwedge_{p,q} (\bar{p} = [p] \wedge \bar{q} = [q] \Rightarrow \neg (p(fv)q)).$$

In view of (1), Conclusion IV.24 and Definition IV.21 we see that there are p_1, p_2, p_3 such that

(5) $$\bar{p} = [p_1] \wedge p_1 \in Vr,$$

(6) $$\bar{p} = [p_3] \wedge \bar{q} = [p_2],$$

(7) $$p_3 \in C_{p_2}.$$

By (4) and (6) we have

(8) $$\neg (p_3(fv)p_2)$$

and by Axiom III.3 and formulas (6) and (5) we obtain

(9) $$p_3 \in Vr.$$

By referring to Definition III.1 and formulas (8), (9) and

(7) we can state that there are s_0, s_1, s_2, s_3 such that

(10) $$s_3 = \rho^{\omega}(s_0, s_1, s_2),$$

(11) $$s_3 \in C_{p_2},$$

(12) $$p_3 \in C_{s_2},$$

(13) $$p_3 \approx s_1.$$

Now let

(14) $$\bar{s} = [s_3] \wedge \bar{s}_0 = [s_0] \wedge \bar{s}_1 = [s_1] \wedge \bar{s}_2 = [s_2].$$

By Theorem IV.28, formulas (14) and (10) yield

(15) $$\bar{s} = \bar{\rho}^{\omega}(\bar{s}_0, \bar{s}_1, \bar{s}_2).$$

It follows from Lemma IV.4 and formulas (11), (12), (6), (13) and (14) that

(16) $$\bar{s} \in \bar{C}_{\bar{q}} \wedge \bar{p} \in \bar{C}_{\bar{s}_2} \wedge \bar{p} = \bar{s}_1.$$

Now (15) and (16) yield a formula which is in contradiction with (2). This concludes the proof.

THEOREM IV.30b

$$\bar{p} \in {}^{k+1}\bar{S} \Leftrightarrow \bar{p} \in {}^{k}\bar{S} \vee \bigvee_{n\geq 1} \bigvee_{\bar{p}_0, \bar{p}_1, \ldots, \bar{p}_n \in {}^{k}\bar{S}}$$

$$(\bar{p} = \bar{\rho}(\bar{p}_0, \bar{p}_1, \ldots, \bar{p}_n) \wedge \bar{p}_0 \notin \bar{O} \wedge \bar{c}^{n+1}(\bar{a}, \bar{a}_1, \ldots, \bar{a}_n, \bar{a}_0)) \vee$$

$$\vee \bigvee_{\bar{p}_0, \bar{p}_1, \bar{p}_2, \bar{p}_3} (\bar{p} = \bar{\rho}^{\omega}(\bar{p}_0, \bar{p}_1, \bar{p}_2) \wedge \bar{p}_2 \in {}^{k}\bar{S} \wedge \bar{p}_3 \overline{(fv)} \bar{p}_2 \wedge$$

$$\wedge \bar{p}_3 = \bar{p}_1 \wedge \bar{c}^3(\bar{a}, \bar{a}_1, \bar{a}_2, \bar{a}_0)).$$

P r o o f. To prove the direct implication we assume that

(1) $$\bar{p} \in {}^{k+1}S.$$

It follows from (1) and Conclusion IV.25 that there is a q_1 such that

(2) $$\bar{p} = [q_1]$$

and

(3) $$q_1 \in {}^{k+1}S.$$

Thus, in view of Definition III.2b and formula (3), one of the following three cases can hold:

(i) $$q_1 \in {}^{k}S,$$

(ii) $$\bigvee_{n\geq 1} \bigvee_{p_0, p_1, \ldots, p_n \in {}^{k}S} (q_1 = \rho(p_0,p_1,\ldots,p_n) \wedge \rho \notin O \wedge$$
$$\wedge\ c^{n+1}(\alpha,\alpha_1,\ldots,\alpha_n,\alpha_0)),$$

(iii) $$\bigvee_{p_0, p_1, p_2, p_3} (q_1 = \rho^{\omega}(p_0,p_1,p_2) \wedge p_2 \in {}^{k}S \wedge p_3(fv)p_2 \wedge$$
$$\wedge\ p_3 \approx p_1 \wedge c^{3}(\alpha,\alpha_1,\alpha_2,\alpha_0)).$$

In case (i), $\bar{p} \in {}^{k}\bar{S}$ in view of (2) and in accordance with Conclusion IV.25, and hence the disjunction which is the consequent of the implication under consideration is true.

In case (ii), the proof of the fact that the consequent of the direct implication being proved is true is modelled on the proof of formula (v) in case (ii) of the proof of Theorem IV.22b. It suffices merely to refer to Conclusion IV.25 instead of Conclusion IV.20 and to Lemma III.1 instead of Lemma II.2, and also to make use of Axiom III.2 and Conclusion IV.23.

Let us accordingly consider case (iii). In view of

Convention II.3c and (iii) we have

(4) $$q_1 = \rho^{\omega}(r_0,r_1,r_2) \wedge r_2 \in {}^{k}S \wedge r_3(fv)r_2 \wedge r_3 \approx r_1 \wedge$$
$$\wedge\ c^3(b_1,c_1,c_2,c_0).$$

It follows from (4) that $q_1=\rho^{\omega}(r_0,r_1,r_2)$. Hence by Definition III.1* we have

(5) $$r_0 \in O \wedge r_1 \in Vr \wedge r_2 \in E.$$

Note that

(6) $$\bar{r}_0 = [r_0] \wedge \bar{r}_1 = [r_1] \wedge \bar{r}_2 = [r_2] \wedge \bar{r}_3 = [r_3].$$

It follows from Conclusions III.1a,b and II.5b, formulas (5) and (3) and Lemma III.1 that $r_0,r_1,r_2,q_1 \in D(\iota)$. Hence by formulas (6) and (2), Theorem IV.16a and Conventions II.3c and IV.3c we have

(7) $$\bar{c}_0 = [c_0] \wedge \bar{c}_1 = [c_1] \wedge \bar{c}_2 = [c_2] \wedge \bar{a} = [b_1].$$

In view of Theorem IV.28 and formulas (6), (2) and (4) we state that

(8) $$\bar{p} = \bar{\rho}^{\omega}(\bar{r}_0,\bar{r}_1,\bar{r}_2).$$

From Conclusion IV.25 and formulas (6) and (4) we obtain

(9) $$\bar{r}_2 \in {}^{k}\bar{S},$$

and in view of Definition IV.29 and formulas (6) and (4) we have

(10) $$\bar{r}_3(\overline{fv})\bar{r}_2.$$

Formulas (4) and (6) yield immediately

(11) $$\bar{r}_3 = \bar{r}_1.$$

Further we conclude from Definition IV.8 and formulas (7) and (4) that

(12) $$\overline{c}^3(\overline{a},\overline{c}_1,\overline{c}_2,\overline{c}_0).$$

Now formulas (8)-(12) taken jointly allow us to state that the consequent of the implication being proved is also true in case (iii). This concludes the proof of the direct implication of Theorem IV.30b.

To prove the converse implication we assume that the disjunction which is the right side of the theorem under consideration is true. Hence one of the following three cases must hold:

(*) $$\overline{p} \in {}^k\overline{S},$$

(**) $$\bigvee_{n\geq 1} \bigvee_{\overline{p}_0,\overline{p}_1,\ldots,\overline{p}_n \in {}^k\overline{S}} (\overline{p} = \overline{p}(\overline{p}_0,\overline{p}_1,\ldots,\overline{p}_n) \wedge \overline{p}_0 \notin \overline{O} \wedge$$
$$\wedge\ \overline{c}^{n+1}(\overline{a},\overline{a}_1,\ldots,\overline{a}_n,\overline{a}_0)),$$

(***) $$\bigvee_{\overline{p}_0,\overline{p}_1,\overline{p}_2,\overline{p}_3} (\overline{p} = \overline{p}^{\omega}(\overline{p}_0,\overline{p}_1,\overline{p}_2) \wedge \overline{p}_2 \in {}^k\overline{S} \wedge \overline{p}_3(\overline{fv})\overline{p}_2 \wedge$$
$$\wedge\ \overline{p}_3 = \overline{p}_1 \wedge \overline{c}^3(\overline{a},\overline{a}_1,\overline{a}_2,\overline{a}_0)).$$

It must be demonstrated that $\overline{p}\in {}^{k+1}\overline{S}$ in each of these three cases.

In case (*) $\overline{p}\in {}^{k+1}\overline{S}$ follows immediately from Conclusion IV.25 and Definition III.2b. The proof of $\overline{p}\in {}^{k+1}\overline{S}$ in case (**) is analogical to the proof of the left side of Theorem IV.22b when we assume the truth of the second constituent of disjunction (v). It suffices to refer to Conclusion IV.25 instead of

Conclusion IV.20 and to Definition III.2b instead of Definition II.4b, and also to make use of Conclusion IV.23.

Consider now case (***). Then we have

(1) $$\overline{p} = \overline{\rho}^{\omega}(\overline{s}_0,\overline{s}_1,\overline{s}_2),$$

(2) $$\overline{s}_2 \in {}^{k}\overline{S},$$

(3) $$\overline{s}_3(\overline{fv})\overline{s}_2,$$

(4) $$\overline{s}_3 = \overline{s}_1,$$

(5) $$\overline{c}^3(\overline{a},\overline{d}_1,\overline{d}_2,\overline{d}_0).$$

By Theorem IV.28 it follows from (1) that there are s_0,s_1,s_2,p_1 such that

(6) $$\overline{s}_0 = [s_0] \wedge \overline{s}_1 = [s_1] \wedge \overline{s}_2 = [s_2]$$

and

(7) $$\overline{p} = [p_1] \wedge p_1 = \rho^{\omega}(s_0,s_1,s_2).$$

By formulas (3) and (2), Definition IV.29, Conclusion IV.25 and Lemma III.2 there are s_4,s_4 such that

(8) $$\overline{s}_2 = [s_4] \wedge \overline{s}_3 = [s_5]$$

and

(9) $$s_5(fv)s_4 \wedge s_4 \in {}^{k}S.$$

It follows from (6) and (8) that $s_2 \approx s_4$, and hence, if we assume that

(10) $$p_2 = \rho^{\omega}(s_0,s_1,s_4),$$

then by (7), Axiom I.1a and Conclusion III.3 we have $p_1 \approx p_2$. Then it follows from (7) that

(11) $$\overline{p} = [p_2].$$

It also follows from (4), (8) and (6) that

(12) $$s_5 \approx s_1.$$

It can easily be seen that the expressions s_0, s_1, s_4, p_2 belong to $D(\iota)$. Hence in accordance with Conventions IV.3c and II.3c and in view of (11), (6), (8) and Theorem IV.16a we have

(13) $$\overline{a} = [a_2] \wedge \overline{d}_0 = [d_0] \wedge \overline{d}_1 = [d_1] \wedge \overline{d}_2 = [d_4].$$

By availing ourselves of Conclusion IV.6 we obtain from (5) and (13)

(14) $$c^3(a_2, d_1, d_4, d_0).$$

By referring to Definition III.2b and formulas (10), (9), (12) and (14) we find that $p_2 \in {}^{k+1}S$, whence by Conclusion IV.25 and formula (11) we have $\overline{p} \in {}^{k+1}\overline{S}$, which was to be demonstrated. The proof has thus been completed.

NOTES

(1) When speaking about languages we mean graphic languages only. A formalized theory of phonic languages (an axiomatic theory of phonology) was presented by T.Batóg [1967].

(2) We interpret here the word "part" in the sense drawn from Stanisław Leśniewski's mereology [1927-31].

(3) The essential outline of **TSCL** was presented in my paper [1981b] read at the 27th Conference on the History of Logic, dedicated to the works of Kazimierz Ajdukiewicz, on October 16, 1981.

(4) See for instance J.Lambek [1961], Y.Bar-Hillel [1964], M.J.Cresswell [1973,1977].

(5) The definition

$$p \in I \Rightarrow (|p| = n \Leftrightarrow p \in {}^{n}I)$$

is dual to Definition I.3.

(6) See for instance R.Montague [1970], M.J.Cresswell [1973,1977], A.Nowaczyk [1978].

(7) We shall later demonstrate (see Conclusion II.9d) that $D(\iota)$ is a non-empty set.

(8) See for instance T.Kubiński [1959], B.Diankov [1974, 1979].

(9) See for instance K.Ajdukiewicz [1935], Y.Bar-Hillel [1953].

(10) Note in connection with this that Axiom II.9 does not guarantee that compound words in $\mathcal{L}$ are its compound

expressions. Hence there may occur in $\mathscr{L}$ such words which have an index in which we are not able unambiguously to distinguish the main functor and its arguments. We mean here expressions which are structurally ambiguous.

(11) These definitions are given later on in this chapter.

(12) See K.Ajdukiewicz [1935].

(13) See for instance M.J.Cresswell [1973], p.71 and [1977] p.258.

(14) Cf. R.Suszko [1958], p.236.

(15) An essential use of these definitions is made in the next chapter. The intuitive sense of the word "constituent" (not very adequate a translation of the term used by Ajdukiewicz) differs from the meaning of the word "part" in Leśniewski's mereology.

(16) Cf. K.Ajdukiewicz [1960] and W.Marciszewski [1977, 1978a].

(17) Not necessarily unambiguous.

(18) We can adopt the convention that the constituent of the zero order of the expression p occupies the position 0.

(19) Cf. J.Słupecki, K.Hałkowska, K.Piróg-Rzepecka [1976] p.278.

(20) Cf. P.Dembiński, J.Małuszyński [1981], pp.11-2.

(21) Examples of trees are to be found further in this section. Those trees have as their elements expression tokens and hence concrete labels (see Figs. II.1-II.6). Tree I of a compound expression p is a graphic illustration of the derivation tree (see for instance J.E.Hopcroft, J.O.Ullman

[1969]) treated as a graph, namely as a graph $(\tilde{C}_p, \longrightarrow)$, where $\tilde{C}_p$ is the set of all those elements of p which are its parts (see Definition II.12), and $\longrightarrow$ is the *relation of direct syntactical subordination* (see K.Ajdukiewicz [1960]), defined by the formula $q \longrightarrow r \Leftrightarrow q,r \in C_p \wedge q \in C_r^1$.

(22) Such an ordering of the elements of tree I is related to the linear ordering of the set of the nodes of the tree $(\tilde{C}_p, \longrightarrow)$, consisting of all those constituents of a given expression p which are its parts; that ordering is done by a relation which might be termed the *relation of an earlier syntactic position*. Its formal, somewhat complicated, definition can easily be formulated if we consider that every element of p which is its part occupies in it a fixed syntactic position.

(23) Such operators are called operators of one argument. They include, e.g., ordinary quantifiers, the description operator, the algebraic summation symbol $\sum$, the integral symbol $\int$, etc. Next to such operators those of n arguments are also being discussed (for $n \geq 1$) (see L.Borkowski [1958, 1970] p.22). For instance quantifiers with restricted range are operators of two arguments.

(24) The symbolic formulation of this definition, which is of little practical use, is omitted here.

(25) Some comments on that matter are to be found in my paper [1983b], pp.308-9.

(26) As we have seen, **TETk** is the core of both the theory **TSCL** of simple categorial languages and the theory **TSCω-L** of

ω-languages.

(27) Conventions concerning the variables which range over the sets $\overline{Lb}$, $\overline{I}$, $\overline{E}$, $2^{\overline{Lb}}$ are given later (see Conventions IV.2a,b, IV.3a-c).

(28) This is why the verbal formulations of induced theorems of dual theories will often be omitted.

(29) For the sake of precision we add to **TETp** the definition $\overline{p}\in\overline{I} \Rightarrow (|\overline{p}|=n \Leftrightarrow \overline{p}\in{}^{n}\overline{I})$, which is induced by the dual analogon of Definition I.3. It defines the concept of the length of an abstract index.

(30) The expansion of the expression $\overline{\rho}(\overline{p}_0,\overline{p}_1,\ldots,\overline{p}_n)$ is the sequence $\overline{p}_0,\overline{p}_1,\ldots,\overline{p}_n$ of word types of which that expression consists.

(31) The expressions $\overline{r}(\overline{p}/\overline{q})^{\overline{n}}\overline{s}$ and $\overline{r}(\overline{p}/\overline{q})\overline{s}$ are read analogically to the corresponding expressions $r(p/q)^{n}s$ and $r(p/q)s$.

REFERENCES

AJDUKIEWICZ, K.

[1934] *W sprawie uniwersaliów (Concerning Universals)*, **Przeglad Filozoficzny**, No.37, pp.219-34.

[1935] *Die syntaktische Konnexität*, **Studia Philosophica**, Leopoli, Vol.1, pp.1-27. English translation: *Syntactic Connection*, in: McCall. S. (ed.), **Polish Logic 1920-1939**, Oxford 1967, pp.207-31.

[1960] *Związki składniowe między członami zdań oznajmujących (Syntactic Relation Between Elements of Declarative Sentences)*, **Studia Filozoficzne**, No.6 (21), pp.73-86. First presented in English at the International Linguistic Symposium in Erfurt, September 27 - October 2, 1958.

BAR-HILLEL, Y.

[1950] *On Syntactical Categories*, **Journal of Symbolic Logic**, Vol.15, pp.1-16. Reprinted in: [1964], pp.19-37.

[1953] *A Quasi-arithmetical Notation for Syntactic Description*, **Language**, Vol.29, pp.47-58. Reprinted in: [1964], pp.61-74.

[1954] *Indexical Expressions*, **Mind**, Vol.63, pp.359-79. Reprinted in: **Aspects of Language**, Jerusalem, pp.69-88.

[1964] **Language and Information**, Selected Essays on their Theory and Application, Addison-Wesley Publishing Co., Reading, Mass..

BAR-HILLEL, Y., GAIFMAN, C., SHAMIR, E.

[1960] *On Categorial and Phrase Structure Grammars*, **Bulletin of the Research Council of Israel**, 9F, pp.1-16. Reprinted in: [1964].

BATÓG, T.

[1967] **The Axiomatic Method in Phonology**, Routlege & Kegan Paul LTD, London.

BOCHEŃSKI, I. M.

[1949] *On the Syntactical Categories*, **The New Schoolmen**, Vol.23, pp.257-80.

BORKOWSKI, L.

[1958] *On Proper Quantifiers*, **Studia Logica**, Vol.VIII, pp. 65-130.

[1970] **Logika Formalna (Formal Logic)**, Warszawa.

BRYLL, G., MIKLOS, S.

[1977] *Teoria konkretnych i abstrakcyjnych słów (A Theory of Concrete and Abstract Words)*, **Publications of the Opole Teachers' Training College Press**, Mathematics Series, Vol.XX, pp.63-76.

BUSZKOWSKI, W.

[1982a] **Lambek's Categorial Grammar**, Adam Mickiewicz University, Poznań.

[1982b] *Compatibility of a Categorial Grammar with an Associated Category Systems*, **Zeitschrift für mathematische Logik und Grundlagen der Mathematik**, Vol.28, pp.229-38.

[1982c] *Some Decision Problems in the Theory of Syntactic Categories*, **Zeitschrift für Mathematische Logik und Grundlagen der Mathematik**, Vol.28, pp.539-48.

[1984] *A Note on the Lambek-Van Benthem Calculus*, **Bulletin of the Section Logic**, Polish Academy of Sciences, No.13, pp.31-37.

[1985] *Algebraic Models of Categorial Grammars*, in: Dorn G., Weingartner P. (eds.), **Foundations of Logic and Linguistics:** Problems and their Solutions, New York.

[1986] *Typed Functorial Languages*, **Bulletin of the Polish Academy of Sciences**, Mathematics, Vol.34, No.7-8, pp. 495-505.

[1987] *Generative Power of Categorial Grammars*, in: Oehrle R., Bach E. and Wheeler D. (eds.), **Categorial Grammars and Natural Language Structures**, D. Reidel, Dordrecht.

BUSZKOWSKI, W., VAN BENTHEM, J., MARCISZEWSKI, W. (eds.)

[1988] **Categorial Grammar**, Benjamins, Amsterdam.

CARNAP, R.

[1937] **The Logical Syntax of Language,** Harcourt & Kegan, New York.

[1942] **Introduction to Semantics**, Cambridge, Mass..

CHOMSKY, N.

[1957] **Syntactic Structures**, Mouton and Co., The Hague.

CRESSWELL, M. J.

[1973] **Logics and Languages**, Mathuen, London.

[1977] *Categorial Languages*, in: Surma S. J. (ed.), On Leśniewski's Systems, **Studia Logica**, Vol.XXXVI, No.4, pp.257-69.

CURRY, H. B.

[1961] *Some Logical Aspects of Grammatical Structure*, in: Jakobson R. (ed.) [1961], pp.57-68.

[1963] **Foundations of Mathematical Logic**, McGraw-Hill, New York.

DEMBIŃSKI, P., MAŁUSZYŃSKI, J.

[1981] **Matematyczne Metody Definiowania Języków Programowania (Mathematical Methods of Defining Programming Languages)**, Warszawa.

DIANKOV, B.

[1974] *An Attempt to Generalization and Extension the Ajdukiewicz's Index Method and Corresponding Metalogical Calculus*, **Ruch Filozoficzny**, Vol.XXXII, No.4, pp.283-88.

[1979] *A New Approach to the Ajdukiewicz's Index Method and the Corresponding Metalogical Calculus*, in: **A Semiotic Landscape** (Proceedings of the First Congress of the International Association for Semiotic Studies, Milan 1974), The Hague-Paris-New York, pp.516-20.

FREGE, G.

[1879] *Begriffsschrift, eine der arithmetischen nachbildete Formelsprache des reinen Denkens*, Halle. Reprinted in: [1964].

[1964] Angelelli I. (ed.), **Begriffsschrift und andere Aufsätze**, Wissenschaftliche Buchgesellschaft/G.Olms, Darmstadt/Hildesheim.

GEACH, P. T.

[1970] *A Program for Syntax*, **Synthese**, Vol.XXII, pp.3-17. Reprinted in: Davison D., Herman H. (eds.), **Semantics of Natural Language**, D.Reidel, Dordrecht.

GROSS, M.,LENTIN, A.

[1967] **Notions sûr les Grammaires Formalles**, Paris.

GRZEGORCZYK, A.

[1981] **An Outline of Mathematical Logic**, Warszawa-Dordrecht-Boston.

HIŻ, H.

[1957] *Types and Environments*, **Philosophy of Sciences**, Vol.24, pp.215-20.

[1960] *The Intuitions of Grammatical Categories*, **Methodos**, Vol.XII, No.48, pp.1-9.

[1961] *Congrammaticality, Batteries of Transformations and Grammatical Categories*, in: Jakobson R. (ed.) [1961], pp.43-50.

[1967] *Grammar Logicism*, **The Monist**, Vol.51, pp.110-27. Reprinted in: Buszkowski W., Van Benthem J., Marciszewski W. (eds.) [1988], pp.263-82.

[1968] *Computable and Uncomputable Elements of Syntax*, in: Van Rootselaar B., Staal J. F. (eds.), **Logic, Methodology and Philosophy Science**, Vol.3, North-Holland Publishing, Amsterdam.

HOPCROFT, J. E., ULLMAN, J. O.

[1969] **Formal Languages and their Relation to Automata,** Addison-Wesley Publishing Co., Reading, Mass..

HUSSERL, E.

[1900-1] **Logische Utersuchungen,** Vol.I, Halle 1900, Vol.II, Halle 1901.

JAKOBSON, R. (ed.)

[1961] **Structure of Language and its Mathematical Aspects,** Proceedings of Symposia in Applied Mathematics, Vol.XII, AMS, Providence, Rhode Island.

KUBIŃSKI, T.

[1958] *On Extension of the Theory of Syntactic Categories,* **Acta Universitatis Wratislaviensis**, No.12, pp.19-36.

LAMBEK, J.

[1958] *The Mathematics of Sentence Structure,* **American Mathematical Monthly**, Vol.65, pp.154-70.

[1961] *On the Calculus of Syntactic Types,* in: Jakobson R. (ed.) [1961], pp.166-78.

LEŚNIEWSKI, S.

[1927-31] *O podstawach matematyki,* **Przegląd Filozoficzny,** Vols.XXX-XXXIV. English translation: *On the Foundations of Mathematics,* Topoi, Vol.II, No.1, pp.7-52.

[1929] *Grundzüge eines neuen Systems der Grundlagen der Mathematik,* **Fundamenta Mathematicae,** Vol.14, pp.1-81.

[1930] *Über die Grundlagen der Ontologie*, Comptes rendus des séances de la Société des Sciences et des Lettres de Varsovie, Classe III, Vol.XXIII, Warszawa, pp.111-32.

LEWIS, D.

[1970] *General Semantics*, **Synthese**, Vol.XII, pp.18-67. Reprinted in: Davidson D., Herman H. (eds.), **Semantics of Natural Language**, D.Reidel, Dordrecht 1972.

ŁUKASIEWICZ, J.

[1929] **Elements of Mathematical Logic**, Warszawa-Oxford 1963 (first published in Polish in 1929).

MARCISZEWSKI, W.

[1977] *Opis syntaktyczny mowy zależnej na gruncie gramatyki kategorialnej (Syntactic Description of Indirect Speech in Categorial Grammar)*, **Studia Semiotyczne**, Vol.VII, pp.206-26.

[1978a] *A Syntactic Description of Oblique Speech in Terms of Categorial Grammar*, **Revista Brasiliera de Linguistica**, Vol.3, No.1.

[1978b] *On Categorial Grammar and Logical Form* (editorial inroduction) in: Marciszewski W.(od.), Studies in Formal Languages, **Studia Logica**, Vol.XXXVII, No.1, pp.1-5.

[1981] (ed.) **Dictonary of Logic as Applied to the Study of Language: Concepts, Methods, Theories**, Martinus Nijhoff, The Hague.

[1989] *Toward Universal Grammars Carnap's and Ajdukiewicz's Contributions*, in: Szaniawski K. (ed.), **The Vienna Circle and Lvov-Warsaw School**, Kluwer Academic Publishers, pp.87-112.

MARKOV, A. A.

[1954] **Teoria Algorifmov**, Izd. Akad. Nauk SSSR, Moskva-Leningrad. English translation: **The Theory of Algorithms**, S.Mouson, Jerusalem 1961.

MONTAGUE, R.

[1970] *Universal Grammar*, **Theoria**, Vol.XXXVI, pp.373-98. Reprinted in: [1974].

[1973] *The Proper Treatment of Quantification in Ordinary English*, in: Hintikka J., Moravicsik J.M.E., Suppes P. (eds.), **Approaches to Natural Language**, D.Reidel, Dordrecht. Reprinted in: [1974].

[1974] Thomason R.H. (ed.), **Formal Philosophy:** Selected Papers of Richard Montague, Yale Univ. Press, New Haven, Conn..

MORRIS, C. S.

[1938] *Foundations of the Theory of Signs*, **International Encyclopedia of Unified Science**, Vol.1, No.2, Univ. of Chicago Press, Chicago, 5th ed. 1947.

[1946] **Signs, Language and Behaviour**, New York.

NOWACZYK, A.

[1978] *Categorial Languages and Variable-Binding Operators*, in: Marciszewski W. (ed.), Studies in Formal Languages, **Studia Logica**, Vol.XXXVII, No.1, pp.27-39.

PEIRCE, CH. S.

[1931-35] Hartshorne C., Weiss P. (eds.), **Collected Papers of Charles Sanders Peirce**, Vols.1-5, Cambridge, Mass..

PELC, J.

[1967] *A Functional Approach to the Logical Semiotics of Natural Language*, in: **Sign, Language, Culture**, The Hague-Paris 1970, pp.89-112 (first published in Polish in 1967).

[1971] **Studies in Functional Logical Semiotics of Natural Languages**, Mouton and Co., The Hague.

SŁUPECKI, J.

[1955] *System logiczny bez operatorów (Logical System Without Operators)*, **Studia Logica**, Vol.III, pp.98-124.

SŁUPECKI, J., HAŁKOWSKA, K., PIRÓG-RZEPECKA, K.

[1978] **Logika i teoria mnogości (Logic and Set Theory)**, Warszawa.

STANOSZ, B., NOWACZYK, A.

[1976] **Logiczne podstawy języka (The Logical Foundations of Language)**, Wrocław-Warszawa.

SUSZKO, R.

[1958] *Syntactic Structure and Semantical Reference*, Part I, **Studia Logica**, Vol.VIII, pp.213-44.

[1960] *Syntactic Structure and Semantical Reference*, Part II, **Studia Logica**, Vol.IX, pp.63-93.

[1964] *O kategoriach syntaktycznych i denotacjach wyrażeń w językach sformalizowanych (Syntactic Categories and Denotation of Expressions in Formalized Languages)*, in: **Rozprawy Logiczne**, Logical Dissertations (to the memory of Kazimierz Ajdukiewicz), Warszawa, pp.193-204.

TARSKI, A.

[1933] *The Concept of Truth in Formalized Languages*, in: [1956] (first published in Polish in 1933).

[1956] **Logic, Semantics, Metamathematics. Papers from 1923 to 1938**, Oxford .

VAN BENTHEM, J.

[1984] *The Logic of Semantics*, in: Landman F., Veltman F. (eds.), **Varietes of Formal Semantics**, Foris, Dordrecht.

[1986] **Essays in Logical Semantics**, D.Reidel, Dordrecht.

VANDERVEKEN, D. R.

[1976] *The Leśniewski-Curry Theory of Syntactical Categories and the Categorially Open Functors*, **Studia Logica**, Vol.XXXV, No.2, pp.191-201.

WYBRANIEC SKARDOWSKA, U.

[1981a] *On the Theory of Label-Tokens*, **Bulletin of the Section of Logic**, Polish Academy of Sciences, Vol.10, No.1, pp.30-4.

[1981b] *Formalizacja teorii języków syntaktycznie kategorialnych (A Formalization of the Theory of Syntactically Categorial Languages)*, Proceedings of the 27th Conference on the History of Logic, Cracow, October 16-18, 1981, pp.21-6.

[1983a] *O teorii napisów egzemplarzy (On the Theory of Label Tokens)*, Publications of the Opole Engineering College, Mathematics Series, No.4, pp.121-5.

[1983b] *O istocie i semantycznych własnościach imion własnych (On the Essence and Semantic Properties of Proper Names)*, **Onomastica**, Polish Academy of Sciences Vol.XXVIII, pp.298-310.

[1986] *On the Type-Token Relationships*, **Bulletin of the Section of Logic**, Polish Academy of Sciences, Vol.15, No.4, pp.164-71.

[1988] *Logiczne podstawy ontologii składni języka (Logical Foundations of Language Syntax Ontology)*, **Studia Filozoficzne**, Logika a Filozofia, No.6-7 (271-272), pp.264-84.

[1989] *On the Eliminatibility of Ideal Linguistic Entities*, **Studia Logica**, Vol.XLVIII, No.4, pp.587-615.

[1990] *O dwóch podejściach do formalizacji teorii napisów (On Two Approaches to the Formalization of the Theory of Labels)*, Publications of the Opole Teachers' Training College Press, Mathematics Series, Vol.XXVII, pp.33-48.

Nijhoff International Philosophy Series

1. N. Rotenstreich: *Philosophy, History and Politics.* Studies in Contemporary English Philosophy of History. 1976 ISBN Pb 90-247-1743-4
2. J.T.J. Srzednicki: *Elements of Social and Political Philosophy.* 1976 ISBN Pb 90-247-1744-2
3. W. Tatarkiewicz: *Analysis of Happiness.* Translated from Polish by E. Rothert and D. Zieliński. 1976 ISBN 90-247-1807-4
4. K. Twardowski: *On the Content and Object of Presentations.* A Psychological Investigation. Translated from Polish with an Introduction by R. Grossmann. 1977 ISBN Pb 90-247-1926-7
5. W. Tatarkiewicz: *A History of Six Ideas.* An Essay in Aesthetics. Translated from Polish a.o. by C. Kasparek. 1980 ISBN 90-247-2233-0
6. H.W. Noonan: *Objects and Identity.* An Examination of the Relative Identity Thesis and Its Consequences. 1980 ISBN 90-247-2292-6
7. L. Crocker: *Positive Liberty.* An Essay in Normative Political Philosophy. 1980 ISBN 90-247-2291-8
8. F. Brentano: *The Theory of Categories.* Translated from German by R.M. Chisholm and N. Guterman. 1981 ISBN 90-247-2302-7
9. W. Marciszewski (ed.): *Dictionary of Logic* as Applied in the Study of Language. Concepts - Methods - Theories. 1981 ISBN 90-247-2123-7
10. I. Ruzsa: *Modal Logic with Descriptions.* 1981 ISBN 90-247-2473-2
11. P. Hoffman: *The Anatomy of Idealism.* Passivity and Activity in Kant, Hegel and Marx. 1982 ISBN 90-247-2708-1
12. M. S. Gram: *Direct Realism.* A Study of Perception. 1983 ISBN 90-247-2870-3
13. J.T.J. Srzednicki, V.F. Rickey and J. Czelakowski (eds.): *Leśniewski's Systems.* Ontology and Mereology. 1984 ISBN 90-247-2879-7
    For 'The Leśniewski Collection' *see also Volume 24.*
14. J. W. Smith: *Reductionism and Cultural Being.* A Philosophical Critique of Sociobiological Reductionism and Physicalist Scientific Unificationism. 1984 ISBN 90-247-2884-3
15. C. Zumbach: *The Transcendent Science.* Kant's Conception of Biological Methodology. 1984 ISBN 90-247-2904-1
16. M. A. Notturno: *Objectivity, Rationality and the Third Realm: Justification and the Grounds of Psychologism.* A Study of Frege and Popper. 1985 ISBN 90-247-2956-4

Volumes 1–8 previously published under the Series Title: Melbourne International Philosophy Series.

Nijhoff International Philosophy Series

17. İ. Dilman (ed.): *Philosophy and Life*. Essays on John Wisdom. 1984 ISBN 90-247-2996-3
18. J. J. Russell: *Analysis and Dialectic*. Studies in the Logic of Foundation Problems. 1984 ISBN 90-247-2990-4
19. G. Currie and A. Musgrave (eds.): *Popper and the Human Sciences*. 1985 ISBN Hb 90-247-2998-X; Pb 90-247-3141-0
20. C. D. Broad: *Ethics*. Lectures given at Cambridge during the Period 1933–34 to 1952–53. Edited by C. Lewy. 1985 ISBN 90-247-3088-0
21. D.A.J. Seargent: *Plurality and Continuity*. An Essay in G.F. Stout's Theory of Universals. 1985 ISBN 90-247-3185-2
22. J.E. Atwell: *Ends and Principles in Kant's Moral Thought*. 1986 ISBN 90-247-3167-4
23. J. Agassi, and I.C. Jarvie (eds.): *Rationality: The Critical View*. 1987 ISBN Hb 90-247-3275-1; Pb 90-247-3455-X
24. J.T.J. Srzednicki and Z. Stachniak: *S. Leśniewski's Lecture Notes in Logic*. 1988 ISBN 90-247-3416-9
    For 'The Leśniewski's Collection' *see also Volume 13.*
25. B.M. Taylor (ed.): *Michael Dummett*. Contributions to Philosophy. 1987 ISBN 90-247-3463-0
26. A. Z. Bar-On: *The Categories and the Principle of Coherence*. Whitehead's Theory of Categories in Historical Perspective. 1987 ISBN 90-247-3478-9
27. B. Dziemidok and P. McCormick (eds.): *On the Aesthetics of Roman Ingarden*. Interpretations and Assessments. 1989 ISBN 0-7923-0071-8
28. J.T.J. Srzednicki (ed.): *Stephan Körner – Philosophical Analysis and Reconstruction*. Contributions to Philosophy. 1987 ISBN 90-247-3543-2
29. F. Brentano: *On the Existence of God*. Lectures given at the Universities of Würzburg and Vienna (1868–1891). Edited and translated from German by Susan F. Krantz. 1987 ISBN 90-247-3538-6
30. Z. Augustynek: *Time. Past, Present and Future*. Essays dedicated to Henryk Mehlberg. (in preparation) ISBN 0-7923-0270-2
31. T. Pawlowski: *Aesthetic Values*. 1989 ISBN 0-7923-0418-7
32. M. Ruse (ed.): *What the Philosophy of Biology Is*. Essays dedicated to David Hull. 1989 ISBN 90–247–3778–8
33. J. Young: *Willing and Unwilling*. A Study in the Philosophy of Arthur Schopenhauer. 1987 ISBN 90-247-3556-4
34. T. Z. Lavine and V. Tejera (eds.): *History and Anti-History in Philosophy*. 1989 ISBN 0-7923-0455-1

Nijhoff International Philosophy Series

35. R.L. Epstein: *The Semantic Foundations of Logic*. Volume 1: Propositional Logics. With the assistance and collaboration of W.A. Carnielli, I.M.L. D'Ottaviano, S. Krajewski and R.D. Maddux. 1990 ISBN 0-7923-0622-8
36. A. Pavković (ed.): *Contemporary Yugoslav Philosophy*. The Analytic Approach. 1988 ISBN 90-247-3776-1
37. A. Winterbourne: *The Ideal and the Real*. An Outline of Kant's Theory of Space, Time and Mathematical Construction. 1988 ISBN 90-247-3774-5
38. K. Szaniawski (ed.): *The Vienna Circle and the Lvov-Warsaw School*. 1989 ISBN 90-247-3798-2
39. G. Priest: *In Contradiction*. A Study of the Transconsistent. 1987 ISBN 90-247-3630-7
40. J. Woleński (ed.): *Kotarbiński: Logic, Semantics and Ontology*. 1990 ISBN 0-7923-0865-4
41. P. Geach (ed.): *Logic and Ethics*. 1991 ISBN 0-7923-1044-6
42. U. Wybraniec-Skardowska: *Theory of Language Syntax*. Categorial Approach. 1991. ISBN 0-7923-1142-6

Further information about our publications on *Philosophy* are available on request.

Kluwer Academic Publishers – Dordrecht / Boston / London